Nanocomposites in Wastewater Treatment

Nanocomposites in Wastewater Treatment

edited by
Ajay Kumar Mishra

PAN STANFORD PUBLISHING

Published by

Pan Stanford Publishing Pte. Ltd.
Penthouse Level, Suntec Tower 3
8 Temasek Boulevard
Singapore 038988

Email: editorial@panstanford.com
Web: www.panstanford.com

British Library Cataloguing in Publication Data
A catalogue record for this book is available from the British Library.

Nanocomposites in Wastewater Treatment
Copyright © 2015 by Pan Stanford Publishing Pte. Ltd.
All rights reserved. This book, or parts thereof, may not be reproduced in any form or by any means, electronic or mechanical, including photocopying, recording or any information storage and retrieval system now known or to be invented, without written permission from the publisher.

For photocopying of material in this volume, please pay a copying fee through the Copyright Clearance Center, Inc., 222 Rosewood Drive, Danvers, MA 01923, USA. In this case permission to photocopy is not required from the publisher.

ISBN 978-981-4463-54-6 (Hardcover)
ISBN 978-981-4463-55-3 (eBook)

Printed in the USA

Contents

Preface ... xiii

1. **Chitosan-Based Polymer Nanocomposites for Heavy Metal Removal** ... 1
   Malathi Sampath, Cross Guevara Kiruba Daniel, Vaishnavi Sureshkumar, Muthusamy Sivakumar, and Sengottuvelan Balasubramanian
   - 1.1 Introduction ... 2
   - 1.2 Why Chitosan? ... 3
   - 1.3 Chitosan-Based Polymer Nanocomposites ... 4
     - 1.3.1 Chitosan Clay Nanocomposite ... 4
     - 1.3.2 Chitosan–Nanoparticle Composite ... 4
   - 1.4 Mechanism of Heavy Metal Removal ... 10
   - 1.5 Concluding Remarks and Future Trends ... 14

2. **Gum-Polysaccharide-Based Nanocomposites for the Treatment of Industrial Effluents** ... 23
   Hemant Mittal, Balbir Singh Kaith, Ajay Kumar Mishra, and Shivani Bhardwaj Mishra
   - 2.1 Introduction ... 24
   - 2.1 Gum Polysaccharides ... 25
     - 2.1.1 Gum Arabic ... 26
     - 2.1.2 Gum Karaya ... 26
     - 2.1.3 Gum Tragacanth ... 27
     - 2.1.4 Gum Xanthan ... 28
     - 2.1.5 Gum Gellan ... 29
     - 2.1.6 Guar gum ... 30
     - 2.1.7 Locust bean gum ... 31
     - 2.1.7 Gum Ghatti ... 32
   - 2.2 Stimuli-Responsive Nanocomposites ... 33
     - 2.2.1 Temperature-Responsive Nanocomposites ... 33
     - 2.2.2 pH-Responsive Nanocomposites ... 35

2.3	Preparation of Nanocomposites		36
	2.3.1 Graft Copolymerization/Cross-Linking		36
	2.3.2 Suspension Polymerization		37
	2.3.3 Polymer Coacervation Process		37
		2.3.3.1 Simple coacervation process	37
		2.3.3.2 Complex coacervation process	38
2.4	Utilization of Nanocomposites for Wasterwater Treatment		38
2.5	Conclusion		39

3. A View on Cellulosic Nanocomposites for Treatment of Wastewater   47

D. Saravana Bavan and G. C. Mohan Kumar

3.1	Introduction	48
3.2	Classification of Natural Fibers	48
3.3	Structure of Natural Fibers	50
3.4	Physical, Mechanical, and Other Properties of Natural Fibers	52
	3.4.1 Problems with Natural Fibers	53
	3.4.2 Limitations of Natural Fibers	53
3.5	Chemical Composition of Natural Fibers	54
	3.5.1 Cellulose	54
	3.5.2 Hemicellulose	55
	3.5.3 Lignin	56
	3.5.4 Pectin and Others	57
3.6	Biocomposites/Green Composites	57
3.7	Wastewater Treatment	58
3.8	Classification of Wastewater Treatment	59
3.9	Dye in Wastewater	61
3.10	Adsorbents in Wastewater	64
	3.10.1 Activated Carbon	65
3.11	Role of Agro-Fibers and Polymers in Handling Wastewater	68
3.12	Biosorption	69
3.13	Activated Carbon from Plant Fibers as Adsorbents	71
3.14	Cellulose Nanocomposite Materials	73

	3.15	Cellulose Nanocrystals (Fibers and Whiskers)	75
	3.16	Conclusion	80

4. Removal of Heavy Metals from Water Using PCL, EVA–Bentonite Nanocomposites — 97

Derrick S. Dlamini, Ajay K. Mishra, and Bhekie B. Mamba

	4.1	Introduction			97
	4.2	Polymeric Nanocomposites			98
		4.2.1	Nanocomposite Formation and Structure		99
			4.2.1.1	Polymer–clay nanocomposite formation	99
			4.2.1.2	Polymer–clay nanocomposite structure	104
	4.3	Polymer–Clay Nanocomposites in Heavy-Metal Removal from Water			109
		4.3.1	Heavy-Metal Adsorption		109
			4.3.1.1	Tailored morphology to enhance adsorption	112
		4.3.2	Heavy-Metal Retention by Granular Filtration		115
		4.3.3	Merits and Limitations of Polymeric Nanocomposites in Water Treatment		118
			4.3.3.1	Merits	118
			4.3.3.2	Limitations	119

5. Role of Polymer Nanocomposites in Wastewater Treatment — 125

Balbir Singh Kaith, Saruchi, Vaneet Thakur, Ajay Kumar Mishra, Shivani Bhardwaj Mishra, and Hemant Mittal

	5.1	Introduction		126
	5.2	Types of Polymer Nanocomposites		128
		5.2.1	Conventional Nanocomposites	128
		5.2.2	Intercalated Nanocomposites	128
		5.2.3	Exfoliated Nanocomposites	129
	5.3	Methods of Preparation		129

		5.3.1	Melt Compounding	129
		5.3.2	In situ Polymerization	129
		5.3.3	Bulk Polymerization	129
		5.3.4	Electrospinning	130
	5.4	Characterization		130
		5.4.1	X-Ray Diffraction	130
		5.4.2	Thermogravimetric Analysis	131
		5.4.3	Transmission Electron Microscopy	132
		5.4.4	Scanning Electron Microscopy	133
	5.5	Application of Polymer Nanocomposites		134
		5.5.1	Dendrimers in Water Treatment	134
		5.5.2	Metal Nanocomposites	135
		5.5.3	Zeolites	135
		5.5.4	Carbonaceous Nanocomposites	136
	5.6	Conclusion		137

6. Nanoparticles for Water Purification — **143**

Pankaj Attri, Rohit Bhatia, Bharti Arora, Jitender Gaur, Ruchita Pal, Arun Lal, Ankit Attri, and Eun Ha Choi

7. Electrochemical Ozone Production for Degradation of Organic Pollutants via Novel Electrodes Coated by Nanocomposite Materials — **167**

Mahmoud Abbasi and Ali Reza Soleymani

	7.1	Introduction		168
	7.2	Ozonation Process in Water and Wastewater Treatment		168
	7.3	Oxidation Mechanism of Ozonation		169
	7.4	Ozone Production Methods		173
		7.4.1	Corona Discharge Method	174
		7.4.2	Photochemical Process	175
		7.4.3	Cold Plasma	175
		7.4.4	Electrochemical Ozone Production	176
	7.5	Anode Materials		178
	7.6	Application of Electrochemically Generated Ozone		184

8. **Core–Shell Nanocomposites for Detection of Heavy Metal Ions in Water** — 191
Sheenam Thatai, Parul Khurana, and Dinesh Kumar

8.1 Introduction — 192
8.2 Classification of Nanocomposites — 193
8.3 Methods for Preparation of Nanomaterials as Nanofillers — 195
    8.3.1 Fe_3O_4 Nanoparticles — 197
    8.3.2 TiO_2 Nanoparticles — 197
    8.3.3 CdS, PbS, and CuS Nanoparticles — 197
    8.3.4 SiO_2 Nanoparticles — 197
8.4 Methods for Preparation of Nanomaterials as Matrix — 198
    8.4.1 Au Nanoparticles — 200
    8.4.2 Ag Nanoparticles — 200
8.5 Methods for Preparation of Nanocomposites — 200
    8.5.1 SiO_2@Ag Core–Shell Nanocomposites — 203
    8.5.2 SiO_2@Au Core–Shell Nanocomposites — 203
    8.5.3 Fe_3O_4@Au Core–Shell Nanocomposites — 204
    8.5.4 Ag@Au Core–Shell Nanocomposites — 205
8.6 Characterization of Nanomaterials and Nanocomposites — 205
    8.6.1 Optical Probe Characterization Techniques — 206
    8.6.2 Electron Probe Characterization Techniques — 206
    8.6.3 Scanning Probe Characterization Technique — 207
    8.6.4 Spectroscopic Characterization Technique — 207
8.7 Sensing and Detection Using Smart Nanocomposites — 208
8.8 Conclusion — 214

9. Conducting Polymer Nanocomposite–Based Membrane for Removal of Escherichia coli and Total Coliforms from Wastewater — 221

Hema Bhandari, Swati Varshney, Amodh Kant Saxena, Vinod Kumar Jain, and Sundeep Kumar Dhawan

- 9.1 Introduction — 222
- 9.2 Development of Polypyrrole-Silver Nanocomposites Impregnated AC Membrane — 226
  - 9.2.1 Synthesis of Ag-NPs — 226
  - 9.2.2 Development of PPY Ag-NPs Impregnated AC Membrane — 226
- 9.3 Antimicrobial Activity Test Methods — 227
  - 9.3.1 Membrane Filtration Method — 228
- 9.4 Characterization of PPY-Ag Nanocomposite — 230
  - 9.4.1 Structural Characterization — 230
    - 9.4.1.1 FTIR spectra — 230
    - 9.4.1.2 Conductivity measurement — 231
    - 9.4.1.3 X-ray diffraction analysis — 231
  - 9.4.2 Thermogravimetric Analysis — 232
  - 9.4.3 Antistatic Study — 234
  - 9.4.4 Morphological Characterization — 236
- 9.5 Antimicrobial Activity — 237
  - 9.5.1 Antimicrobial Mechanism of PPY-Ag Nanocomposite Impregnated AC Fiber — 241
- 9.6 Conclusion — 242

10. Titanium Dioxide–Based Materials for Photocatalytic Conversion of Water Pollutants — 247

Sónia A. C. Carabineiro, Adrián M. T. Silva, Cláudia G. Silva, Ricardo A. Segundo, Goran Dražić, José L. Figueiredo, and Joaquim L. Faria

- 10.1 Introduction — 248
- 10.2 Experiments — 250
  - 10.2.1 Preparation of Titanium Dioxide Supports — 250
  - 10.2.2 Gold Loading — 251
  - 10.2.3 Characterization Techniques — 251

	10.2.4 Catalytic Tests	252
10.3	Results and Discussion	253
	10.3.1 Characterization of TiO_2 Materials	253
	10.3.2 Characterization of Au/TiO_2 Materials	256
	10.3.3 Catalytic Results for DP Photodegradation	258
	10.3.4 Photocatalytic Degradation of Phenolic Compounds using P25 Catalyst	261
10.4	Conclusion	264

Index 271

Preface

A composite is defined as a combination of two or more materials with different physical and chemical properties and distinguishable interface. There are many advantages of composites over many metal compounds, such as high toughness, high specific stiffness, high specific strength, gas barrier characteristics, flame retardancy, corrosion resistance, low density, and thermal insulation. Composite materials are composed of two phases: the continuous phase known as matrix and the dispersed phase known as reinforced materials. Nanomaterials, in particular nanocomposites, have diversified applications in different areas such as biological sciences, drug delivery systems, and wastewater treatment. In nanocomposites, the nanoparticles were incorporated within different functionalized materials such as multiwalled carbon nanotubes, activated carbon, reduced grapheme oxide, and different polymeric matrices.

Water pollution is mainly caused by the pollutants that result in severe environmental problems. In recent years, various methods for heavy metal detection from water have been extensively studied. A different variety of core–shell nanocomposites such as SiO_2@Au and SiO_2@Ag were also used as a tool for water purification. These nanocomposites provide high surface area and a specific affinity for heavy metal adsorption from aqueous systems. The adsorption of different pollutants such as heavy metal ions and dyes from the contaminated water using nanocomposites has attracted significant attraction due to their characteristic properties such as extremely small size, very large surface area, absence of internal diffusion resistance, and high surface-area-to-volume ratio. Metal oxide nanoparticles, including aluminum oxides, titanium oxides, magnesium oxides, cerium oxides, and ferric oxides, have been proved to be very efficient for the removal of various pollutants from the aqueous water.

Nanocomposites have better adsorption capacity, selectivity, and stability than nanoparticles. Magnetic nanocomposites are also a very efficient class of nanocomposites in which magnetic nanoparticles have been used as the reinforcing material. They have the advantages of both magnetic separation techniques and nano-sized materials, which can be easily recovered or manipulated with an external magnetic field. They are also very effective for the removal of both organic and inorganic pollutants from the pollutant water.

This book describes the applications of nanocomposites in various areas, including environmental science, such as remediation and speciation, water research, medicine, and sensors. The application of nanocomposites in wastewater research, which includes organic, inorganic, and microbial pollutants, has also gained more attention in research. The book contains a comprehensive discussion about wastewater research.

Researchers working in the similar domain of research will benefit from the fundamental concepts and advanced approaches described in the book. Researchers involved in the environmental and water research on nanocomposites and their applications will be major beneficiaries of the content of the book. The book will also be beneficial to the researchers who are working for their graduate and postgraduate degrees in the area of nanotechnology. It provides a platform for all researchers as it covers a vast background for the recent literature, abbreviations, and summaries. It will be a worthy read for the researchers in the fields of nanotechnology and engineered materials who are interested in nanocomposites.

The book covers a broader research area of chemistry, physics, materials science, polymer science and engineering, and nanotechnology to present an interdisciplinary approach. It presents the fundamental knowledge with the recent advancements in the research and development of nanocomposites. It discusses the recent approach and prospects about the current research and development in nanocomposites.

Ajay Kumar Mishra

Chapter 1

Chitosan-Based Polymer Nanocomposites for Heavy Metal Removal

Malathi Sampath,[a,*] Cross Guevara Kiruba Daniel,[b,*] Vaishnavi Sureshkumar,[b] Muthusamy Sivakumar,[b] and Sengottuvelan Balasubramanian[a]

[a]*Department of Inorganic Chemistry, Guindy Campus, University of Madras, Chennai 600025, India*
[b]*Division of Nanoscience and Technology, Anna University, BIT campus, Tiruchirappalli 620024, India*
maluscientist2015@gmail.com

New functional nanocomposite materials with impregnated nanoparticles are in the forefront of nanobased water treatment methods. In this chapter, the recent developments on chitosan-based nanocomposites having nanoparticles as impregnated materials and their applications in heavy metal removal will be discussed. More insight can be derived from our work on chitosan-based nanocomposites, which are one of the less expensive nanoproducts developed. The synergistic action of individual polymer (chitosan),

---

*Contributed equally.

Nanocomposites in Wastewater Treatment
Edited by Ajay Kumar Mishra
Copyright © 2015 Pan Stanford Publishing Pte. Ltd.
ISBN 978-981-4463-54-6 (Hardcover), 978-981-4463-55-3 (eBook)
www.panstanford.com

which has inherent properties of heavy metal removal, with metal nanoparticles has been utilized. The hybrid composites exhibit better efficiency than those of conventional materials/products available in market for heavy metal removal.

1.1 Introduction

Industrialization has led to an increase in the release of toxic effluents, including toxic chemicals such as heavy metals. Some of the most common heavy metals are lead, nickel, copper, mercury, chromium, cadmium, and arsenic, which are being released by dental operation, textile, tanning, electroplating, and the paper and pulp industry (Volesky, 2001). These heavy metals cause serious environmental problems by entering into the food chain leading to severe health disorders in humans who are at the top of the food chain (Kortenkamp et al., 1996). There is an urgent need to safeguard water and food resources from heavy metals. It has also become essential to purify water contaminated by heavy metal ions. A large number of strategies are available currently to decontaminate water to make it potable, including reverse osmosis (Monser et al., 2002; Kongsrichroen et al., 2004) ion exchange (Hafez et al., 2004; Modrzejewska et al., 1999), cyanide treatment (Rengaraj et al., 2008), electrochemical precipitation (Rengaraj et al., 2008), adsorption (Zhang et al., 2010; Mohan et al., 2005; Sharma et al., 2008; Hu et al., 2005; Aydin et al., 2009; Zhao et al., 2011; Sawalha et al., 2005; Parsons et al., 2002). Even though reverse osmosis has been successful in effectively reducing heavy metal ions, it is limited by its high operational cost and limited pH range (Wan Nagah et al.). Most of the existing heavy metal removal technologies are limited due to their high cost. It has become necessary to evolve new technologies using widely available low cost materials for the removal of heavy metal ions. Of all the methods, adsorption is the most effective and widely used technique (Rao et al., 2002) due to the usage of commonly available by-products and low cost alternatives such as fly ash (Rao et al., 2002), agricultural wastes (Marshall et al., 1996; Wafoyo et al., 1999), banana pith (Low et al., 1995), chitin and chitosan (Sankararamakrishnan, 2006; Cheung et al., 2003; Joen et al., 2004). The use of nanocomposites of chitosan as an effective adsorbent will be discussed in this chapter. Chitosan is a wonderful biopolymer from chitin, which is widely available in nature. The heavy metal adsorption property of chitosan is increased by additives like nanoparticles.

1.2 Why Chitosan?

Chitosan is a biopolymer obtained from chitin by deacetylation. Chitin is available in nature in abundance and is the second most available biopolymer after cellulose. It is found as a major structural component of mollusks, insects, crustaceans, fungi, algae, and marine invertebrates such as crab, shrimp, shellfish, and krill (Deshpande et al., 1986; Chen et al., 1994; Illyina et al., 1995). Due to its natural origin, chitosan is biodegradable, biocompatible, and antibacterial (Srinivasa et al., 2009), which makes it the most suitable candidate for use as an adsorbent. Existing water treatment technologies have several disadvantages such as being expensive, generating secondary pollutants in sludge, and ineffective in treating effluents with low metal concentrations (Han et al., 2006). Commercially available adsorbents such as activated carbon are highly efficient but are expensive (Cybelle et al., 2011). Throughout the world, the solid waste from the processing of shellfish, krill, shrimps, and crabs constitutes large amount of chitinaceous waste (Nomanbhay et al., 2005), which can be converted to chitosan by partial deacetylation as low cost adsorbent. Chitosan has the highest sorption capacity for several metal ions among the various biopolymers, possibly due to the presence of primary amine at C-2 position of the glucosamine residues (Yi et al., 2005). But it has drawbacks such as softness, tendency to agglomerate or form gels, and nonavailability of reactive binding sites (Nomanbhay et al., 2005). However, the addition of nanoparticles increases the usability of this biopolymer.

Chemical structure of chitin and chitosan.

1.3 Chitosan-Based Polymer Nanocomposites

1.3.1 Chitosan Clay Nanocomposite

Chitosan nanoparticles are being used as adsorbents for heavy metal removal. Recent investigation has focused on heavy metal removal by chitosan nanoparticles with clays such as bentonite, kaolinite, and montmorillonite. Clays have inherent capability to remove heavy metals just like chitin and chitosan. Investigations on nano-chitosan–clay composite for metal ion removal have been reported in the recent period. Khedr et al. have reported the removal of lead by means of modified chitosan–montmorillonite nanocomposite. Nickel (II) removal by chitosan-coated bentonite up to 88% in fixed-bed column has been evaluated (Futalan et al., 2011). Chitosan–clay nanocomposite using montmorillonite has been used for the removal of hexavalent chromium from aqueous solution (Pandey et al., 2011).

1.3.2 Chitosan–Nanoparticle Composite

Chitosan with metal nanoparticle composite has been utilized for different heavy metal removal due to the synergistic action of chitosan and nanoparticles for heavy metal adsorption. Magnetite nanoparticles are found to have heavy metal adsorption capability. Chitosan is known to adsorb a number of heavy metal ions due to the chelation of the amide group of glucosamine. The pH plays a crucial role in the adsorption efficiency of chitosan nanocomposite. As the pH is slowly increased from 2 to 4, there is an increase in the adsorption of heavy metals, but above pH 4, there is a reduction in the adsorption of heavy metals. In one of the reported studies (Liu et al., 2009), magnetic chitosan was used for the removal of Pb^{2+}, Hg^{2+}, Cd^{2+}, Cu^{2+}, and Ni^{2+} by an external magnetic field. The magnetic chitosan, after having adsorbed the heavy metals, can be made to release the metal ions when subjected to ultrasound radiation in weakly acidic deionized water. Thus it is possible to recycle and reuse the magnetic nanocomposite.

Chitosan–magnetite nanocomposites are reported (Namdeo et al., 2008) for the removal of Fe (III) from aqueous solution. Magnetite nanoparticles were held in between chitosan, and the removal of Fe (III) follows Langmuir and Freundlich isotherms. Though there are a number of reports on the adsorption of radioactive ions by chitosan and its derivatives (Akkaya et al., 2008; Anirudhan et al., 2010; Humelnicu et al., 2011; Muzzarelli, 2011), only very few reports are available on the usage of chitosan-based nanocomposites. Ethylenediamine-modified magnetic chitosan particles were reported to adsorb radioactive uranyl ions by Wang et al., 2011. Hritcu et al. (2012) have reported the adsorption of thorium and uranyl ions by unmodified magnetic chitosan particles.

Table 1.1 Chitosan nanocomposites and heavy metal adsorption

S. no.	Composite	Heavy metal	Type	Reference
1.	Chitosan–magnetite nanocomposites	Fe(III)	Adsorption	Namdeo et al. (2008)
2.	Chitosan/poly (acrylicacid) magnetic composite	Cu(II)	Adsorption	Yan et al. (2012)
3.	Multiwalled carbon nanotubes/chitosan Nanocomposite	Ni(II), Cu(II), Cd(II), Zn(II)	Adsorption	Salem et al. (2011)
4.	Magnetic chitosan nanoparticles	Cu(II)	Adsorption	Yuwei et al. (2011)
5.	Hydrogels of chitosan, itaconic, and methacrylic acid	Cu(II)	Adsorption	Milosavljević et al. (2011)
6.	Magnetic chitosan modified with diethylenetriamine	Cu(II), Zn(II), Cr(VI)	Adsorption	Li et al. (2011)
7.	Chitosan-coated sand	Cu(II), Pb(II)	Adsorption	Wan et al. (2010)

(Continued)

Table 1.1 (*Continued*)

S. no.	Composite	Heavy metal	Type	Reference
8.	Chitosan immobilized on bentonite	Cu(II)	Adsorption	Futalan et al. (2011)
9.	Chitosan entrapped in polyacrylamide hydrogel	Pb(II), UO_2(II), Th(IV)	Adsorption	Akkaya et al. (2000)
10.	Chitosan-coated fly ash composite as biosorbent	Cr(VI)	Adsorption	Wen et al. (2011)
11.	Chitosan/activated clay composites	Na(I), K(I), Pb(II)	Adsorption	Eloussaief et al. (2010)
12.	Cross-linked chitosan	U(VI)	Adsorption	Wang et al. (2009)
13.	Chitosan benzoyl thiourea	Co(II), Eu(III)	Adsorption	Metwally et al. (2009)
14.	Quaternary ammonium salt of chitosan	Cr(VI)	Adsorption	Spinelli et al. (2004)
15.	Chitosan-coated onto nonporous ceramic alumina	Cr(VI)	Adsorption	Boddu et al. (2003)
16.	Chitosan beads	As(V)	Adsorption	Chen et al. (2006)
17.	Chitosan flakes	Cr(VI)	Adsorption	Kwok et al. (2009)
18.	Molybdate-impregnated chitosan gel beads	As(V)	Adsorption	Dambies et al. (2000)
19.	Chitosan	Al(III), VO_4(III)	Adsorption	Septhum et al. (2007)
20.	Chitosan coated onto polyvinyl chloride (PVC) beads	Ni(II)	Adsorption	Krishnapriya et al. (2009)

S. no.	Composite	Heavy metal	Type	Reference
21.	Amido-grafted chitosan	Cr(VI)	Adsorption	Kyzas et al. (2009)
22.	Chitosan obtained from silk worm chrysalides	Pb(II)	Adsorption	Paulino et al. (2008)
23.	Aminated chitosan beads	Hg(II)	Adsorption	Jeon et al. (2003)
24.	Chitin	Cd(II)	Adsorption	Benguella et al. (2002)
25.	Cotton fiber/chitosan	Au(III)	Adsorption	Qu et al. (2009)
26.	Chitosan/poly (vinyl alcohol)	Pb(II)	Adsorption	Fajardo et al. (2012)
27.	Chitosan/poly (vinyl alcohol)	Cu(II)	Adsorption	Salehi et al. (2012)
28.	Chitosan-bound Fe_3O_4	Cu(II)	Adsorption	Chang et al. (2005)
29.	Chitosan/attapulgite	Cr(III), Fe(III)	Adsorption	Zou et al. (2011)
30.	N,O-carboxymethyl chitosan/cellulose acetate blend nanofiltration membrane	Cr(III), Cu(II)	Adsorption	Boricha et al. (2010)
31.	Chitin/cellulose	Hg(II)	Adsorption	Tang et al. (2011)
32.	Magnetically modified chitosan	Zn(II)	Adsorption	Fan et al. (2011)
33.	Magnetically modified chitosan	Cr(III)	Adsorption	Geng et al. (2009)
34.	Chitosan cross-linked with a barbital	Hg(II), $CH_3Hg(II)$, $C_6H_5Hg(II)$	Adsorption	Kushwaha et al. (2011)

(*Continued*)

Table 1.1 (*Continued*)

S. no.	Composite	Heavy metal	Type	Reference
35.	Fe–chitosan	Cr(VI)	Adsorption	Zimmermann et al. (2010)
36.	AgNPs/CT membrane	As(III)	Adsorption	Prakash et al. (2012)
37.	Natural and cross-linked chitosan films	Cu(II), Hg(II), Cr(III)	Adsorption	Vieira et al. (2011)
38.	Chitosan-based hydrogels	Pb(II), Cd(II), Cu(II)	Adsorption	Paulino et al (2011)
39.	Chitosan/montmorillonite	Cr(VI)	Adsorption	Fan et al. (2006)
40.	Chitosan/PVC	Cu(II), Ni(II)	Adsorption	Popuri et al. (2009)
41.	Chitosan-coated fly ash composite as biosorbent	Cr(VI)	Adsorption	Wen et al. (2011)
42.	Chitosan-capped gold nanocomposite	Zn(II), Cu(II), Pb(II)	Adsorption	Sugunan et al. (2005)
43.	Chitosan–tripolyphosphate	UO_2(II)	Adsorption	Sureshkumar et al. (2010)
44.	Chitosan immobilized on bentonite	Cu(II)	Adsorption	Futulan et al. (2011)
45.	Chitosan/triethanolamine composites	Ag(I)	Adsorption	Zhang et al. (2012)
46.	Thiourea-modified magnetic ion-imprinted chitosan/TiO_2 composite	Cd(II)	Adsorption	Chen et al. (2012)

Chitin/chitosan nanohydroxyapatite composites have been studied for the removal of Cu(II) by Rajiv Gandhi et al. (2011). It has been found that the sorption capacity of chitosan nanohydroxyapatite is better than that of chitin nanohydroxyapatite and that the selectivity of metal ions by the sorbents was found to vary in the following order: Fe(III) > Cu(II) > Cr(VI). Abou El-Reash et al. (2011) have reported the use of cross-linked magnetic chitosan anthranilic acid glutaraldehyde Schiff's base (CAGS) for the removal of As(V) and Cr(VI). In this study, anion exchange resin has been used along with cross-linked magnetic chitosan for heavy metal removal.

Graphene oxide–chitosan nanocomposite has been utilized for Au(III) and Pd(II) removal (Liu et al., 2012). Graphene oxide, in its graphitic backbone, has a large number of oxygen atoms in the form of epoxy, hydroxy, and carboxyl groups protruding from its layers. Heavy metal adsorption data is usually interpreted by the following equations: Langmuir (Langmuir et al., 1918) and Freundlich (Freundlich et al., 1906) isotherm equations.

Liu et al. (2012) have reported that Langmuir isotherm assumes monolayer adsorption, and the adsorption occurs at specific homogeneous adsorption sites. The expression of the Langmuir isotherm is given by Eq. 1.1.

$$q_e = \frac{q_m b C_e}{1 + b C_e} \qquad (1.1)$$

where q_e (mg/g) is the adsorption capacity of Au(III) and Pd(II) at equilibrium, q_m (mg/g) is the theoretical saturation adsorption capacity for monolayer coverage, C_e (mg/L) is the concentration of Au(III) and Pd(II) at equilibrium, and b (1/mg) is the Langmuir constant related to the affinity of binding sites and is a measure of the energy of adsorption. These constants can be calculated from the slope and intercept of the linear plot of C_e/q_e versus C_e. In addition, the parameter R_L, called the equilibrium parameter, is calculated from Eq. 1.2 to identify whether an adsorption system is favorable or unfavorable.

$$R_L = \frac{1}{1 + b C_0}, \qquad (1.2)$$

where C_0 (mg/g) is the initial concentration. If $R_L > 1$, the adsorption process is favorable; if $R_L = 1$, the process is linear; if $R_L < 1$, the process is unfavorable; and if $R_L = 0$, the process is irreversible.

Different from Langmuir isotherm, the Freundlich isotherm assumes heterogeneous adsorption due to the diversity of the adsorption sites or the diverse nature of the metal ions adsorbed. The Freundlich adsorption equation is given by Eq. 1.3.

$$q_e = kC_e^{1/n} \tag{1.3}$$

where k and n are the Freundlich constants, related to the adsorption capacity of the adsorbent and the adsorption intensity, respectively. One can obtain k and n from the slope and intercept of the linear plot of log q_e versus log C_e.

Similarly, water-dispersible magnetite chitosan/graphene oxide composites have been studied (Fan et al., 2012) for the removal of Pb(II), which is a highly pH-dependent sorption. In our study (unpublished data), a chitosan–magnetite hybrid strip for heavy metal removal was fabricated. The strip was prepared by a simple process of casting technique (Fig. 1.1) and characterized by a field emission scanning electron microscope (Fig. 1.2). Magnetite nanoparticles were initially synthesized by chemical co-precipitation technique and mixed with unmodified chitosan. The strip was then utilized for the removal of Ni^{2+}, Cu^{2+}, and Pb^{2+} from aqueous solution (Figs. 1.3–1.5).

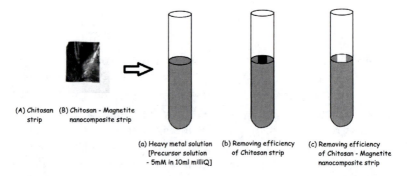

Figure 1.1   Experimental setup for the evaluation of heavy metal removal by chitosan and chitosan–magnetite nanocomposite strip (1 cm × 1 cm).

1.4 Mechanism of Heavy Metal Removal

Adsorption strongly depends on surface area and pore structure of the adsorbent (Kuchta et al., 2005). Metal ion adsorption is strongly

Mechanism of Heavy Metal Removal | 11

Figure 1.2 Field emission scanning electron micrograph of magnetite–chitosan hybrid strip showing clusters of magnetite nanoparticles (size: 30–40 nm) in chitosan matrix.

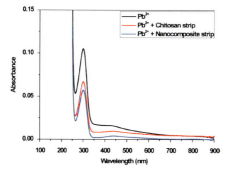

Figure 1.3 UV-visible spectral studies exhibiting the removal of Pb^{2+} by chitosan and hybrid chitosan–magnetite nanoparticle strip.

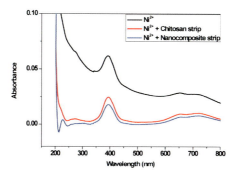

Figure 1.4 UV-visible spectral studies exhibiting the removal of Ni^{2+} by chitosan and hybrid chitosan–magnetite nanoparticle strip.

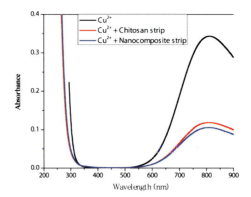

Figure 1.5 UV-visible spectral studies exhibiting the removal of Cu^{2+} by chitosan and hybrid chitosan-magnetite nanoparticle strip.

influenced by the surface modification of the sorbents (Hyung et al., 2009). It may also depend on the ion exchange process; for example, in hydroxyapatite/chitosan nanocomposite, removal of copper is governed by ion exchange in which metal ions in solution replaces Ca^{2+} (Zheng et al., 2007). Also metal ions are sorbed over PO_4H^- (Annadurai et al., 2001). In addition, the presence of metal nanoparticles with chitosan brings in sorption by chelation (Liu et al., 2009) as was the case of magnetite in magnetic chitosan nanocomposite. Chitosan-based nanocomposites are more efficient than chitin-based nanocomposites due to the presence of more number of chelating reactive amino groups in chitosan rather than acetamide groups of chitin (Rajiv Gandhi et al., 2011).

In case of cross-linked magnetic CAGS, the resin was formed as a thin film over magnetite nanoparticles and was reported for the removal of As(V) and Cr(VI) (Abou El-Reash et al., 2011). Functionally free $-NH_2$ groups and the $-C=N$ of Schiff's base group in cross-linked magnetic CAGS were considered active sites for the adsorption of As(V) and Cr(VI). The pH of the solution influences these $-NH_2$ and $-C=N$ groups in cross-linked magnetic CAGS to undergo protonation to $-NH_3^+$ and $-C=N$ and hence the extent of protonation will depend on the solution pH.

The adsorption mechanisms of Au(III) and Pd(II) with graphene oxide chitosan nanocomposite may be mainly due to the ion interaction between protonated amines and tetrachloroaurate, chloropalladate complexes and the coordination of Au(III) and Pd(II) with nitrogen atoms and oxygen atoms. In acid solutions, the amine groups are easily protonated. The pK_a of chitosan ranges from 6.3 to 7.2, and at pH 4, the amine groups are almost 100% protonated. Such protonated amine groups may cause electrostatic interaction of anionic metal complexes, which result from metal chelation by chloride ligands in the present study.

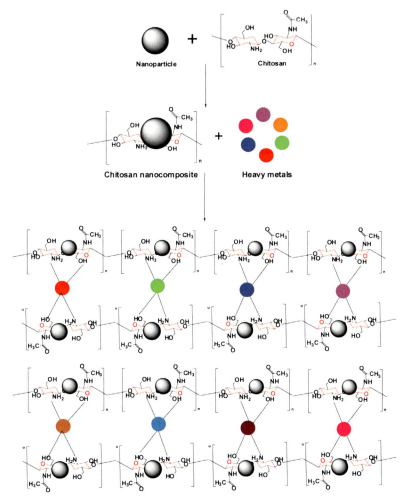

Schematic representation explaining the mechanism of heavy metal adsorption on chitosan-based nanocomposites.

1.5 Concluding Remarks and Future Trends

Heavy metal toxicity is one of the core issues of environmental problems faced by both developed and developing countries. Low cost adsorbents are being developed for heavy metal adsorption and subsequent removal. Biopolymers are being preferred due to their natural source and abundance. Chitosan is one of the wonderful biopolymers available in nature, which is employed for various applications and is available in huge quantity as wastes from crab and shrimp industries. Chitosan has an inherent nature of metal chelation, which makes it an ideal candidate for fabrication of polymer nanocomposite with enhanced performance. In this chapter, we have provided a list of chitosan-based nanocomposites utilized for heavy metal removal and their mechanism of action. We have also indicated our chitosan-based hybrid strip as a case study for enhanced action. Multitasking chitosan-based nanocomposites would be the ideal candidates in near future, which can sense and remove metal ions, bacteria, virus-like microorganisms and other harmful moieties from the environment.

Chemically functionalized chitosan-based nanocomposite films and strips that can clean textile and tannery effluents are some of our research interests being currently pursued.

Acknowledgment

Malathi would like to thank ICMR, India, for Senior Research Fellowship. S.C.G. Kiruba Daniel would like to acknowledge that this work has been catalysed by the RFRS fund provided by the Tamil Nadu State Council of Science and Technology (TNSCST), Government of Tamilnadu.

References

Abou El-Reash, Y.G., Otto, M., Kenawy, I.M., Ouf, A.M. (2011). Adsorption of Cr(VI) and As(V) ions modified magnetic chitosan chelating resin. *Int. J. Biol. Macromol.*, **49**, pp. 513–522.

Akkaya, R., Ulusoy, U. (2008). Adsorptive features of chitosan entrapped in polyacrylamide hydrogel for Pb^{2+}, UO_2^{2+}, and Th^{4+}. *J. Hazard. Mater.*, **151**, pp. 380–388.

Anirudhan, T.S., Rijith, S., Tharun, A.R. (2010). Adsorptive removal of thorium(IV) from aqueous solutions using poly(methacrylic acid)-

grafted chitosan/bentonite composite matrix: Process design and equilibrium studies, *Colloids Surf., A. Phsicochem. Eng. Aspects,* **368**, pp. 13–22.

Annadurai, G., Juang, R.S., Lee, D.J. (2001). Adsorption of heavy metals from water using banana and orange peels, *Water Sci. Technol.,* **47**, pp. 185–190.

Aydln, Y.A., Aksoy, N.D. (2009). Adsorption of chromium on chitosan: Optimization, kinetics and thermodynamics. *Chem. Eng. J.,* **151**, pp. 188–194.

Benguella, B., Benaissa, H. (2002). Cadmium removal from aqueous solutions by chitin: Kinetic and equilibrium studies, *Water Res.,* **36**, pp. 2463–2474.

Boddu, V.M., Aburri, K., Talmott, J., Smith, E. (2003). Removal of hexavalent chromium from wastewater using a new composite chitosan biosorbent, *Environ. Sci. Technol.,* **37**, pp. 4449–4456.

Boricha, A.G., Murthy, Z.V.P. (2010). Preparation of N,O-carboxymethyl chitosan/cellulose acetate blend nanofiltration membrane and testing its performance in treating industrial wastewater, *Chem. Eng. J.,* **157**, pp. 393–400.

Chang, Y-C., Chang, S.W., Chen, D.H. (2006). Magnetic chitosan nanoparticles: Studies on chitosan binding and adsorption of Co(II) ions, *React. Funct. Polym.,* **66,** 335–341.

Chang, Y-C., Chen, D-H. (2005). Preparation and adsorption properties of monodisperse chitosan-bound Fe_3O_4 magnetic nanoparticles for removal of Cu(II) ions, *J. Colloid. Interface Sci.,* **283**, pp. 446–451.

Chen, A., Zeng, G., Chen, G., Hu, X., Yan, M., Guan, S., Shang, C., Lu, L., Zou, Z., Xie, G. (2012). Novel thiourea-modified magnetic ion-imprinted chitosan/TiO_2 composite for simultaneous removal of cadmium and 2,4-dichlorophenol, *Chem. Eng. J.,* **191**, pp. 85–94.

Chen, C.C., Chung, Y.C. (2006). Arsenic removal using a biopolymer chitosan sorbent, *J. Environ. Sci. Health Part A: Toxic/Hazard. Subst. Environ. Eng.,* **41**, pp. 645–658.

Chen, J.P., Chang, K.C. (1994). Immobilization of chitinase on a reversibly soluble-insoluble polymer for chitin hydrolysis, *J. Chem. Technol. Biotechnol.,* **60**, pp. 133–140.

Cheung, W.H., Ng, J.C.Y., Mckay, G. (2003). Kinetic analysis of the sorption of copper ions on chistosan, *J. Chem. Technol. Biotechnol.,* **78**, pp. 562–571.

Dambies, L., Roze, A., Guibal, E. (2000). As(V) sorption on molybdate-impregnated chitosan gel beads (MICB), *Adv. Chitin Sci.*, **4**, pp. 302–309.

Deshpande, M.V. (1986). Enzymatic degradation of chitin and its biological application, *J. Sci. Ind. Res.*, **45**, pp. 277–281.

Eloussaief, M., Benzina, M. (2010). Efficiency of natural and acid-activated clays in the removal of Pb(II) from aqueous solutions, *J. Hazard. Mater.*, **178**, pp. 753–757.

Fajardo, A.R., Lopes, L.C., Rubira, A.F., Muniz, E.C. (2012). Development and application of chitosan/poly(vinyl alcohol) films for removal and recovery of Pb(II), *Chem. Eng. J.*, **183**, pp. 253–260.

Fan, D., Zhu, X., Xu, M., Yan, J. (2006). Adsorption properties of chromium (VI) by chitosan-coated montmorillonite, *J. Biol. Sci.*, **6**, pp. 941–945.

Fan, L., Luo, C., Lv, Z., Lu, F., Qiu, H. (2011). Preparation of magnetic modified chitosan and adsorption of Zn^{2+} from aqueous solutions, *Colloids Surf. B. Biointerfaces*, **88**, pp. 574–581.

Fan, L., Luo, C., Sun, M., Li, X., Qiu, H. (2012). Highly selective adsorption of lead ions by water-dispersible magnetic chitosan/graphene oxide composites, *Colloids Surf. B. Biointerfaces*, **103C**, pp. 523–529.

Freundlich, H.M.F. (1906). Uber die adsorption in losungen, *Z. Phys. Chem.*, **57**, pp. 385–471.

Futalan, C.M., Kan, C.C., Dalida, M.L., Pascua, C., Wan, M.W. (2011). Fixed-bed column studies on the removal of copper using chitosan immobilized on bentonite, *Carbohyd. Polym.*, **83**, pp. 697–704.

Geng, B., Jin. Z.H., Li, T.L., Qi, X.H. (2009). Kinetics of hexavalent chromium removal from water by chitosan-Fe0 nanoparticles, *Chemosphere*, **75**, pp. 825–830.

George, Z.K., Margaritis, K., Nikolaos K.L. (2009). Copper and chromium(VI) removal by chitosan derivatives: Equilibrium and kinetic studies, *Chem. Eng. J.*, **152**, pp. 440–448.

Hafez, A., El-Mariharawy, S. (2004). Design and performance of the two-stage/two pass RO membrane system for chromium removal from tannery water, *Desalination*, **165**, pp. 141–151.

Haibo, L., Shaodan, B., Long, L., Weifang, D., Xin, W. (2011). Separation and accumulation of Cu(II), Zn(II) and Cr(VI) from aqueous solution by magnetic chitosan modified with diethylenetriamine, *Desalination*, **278**, pp. 397–404.

Han, R., Zou, W., Li, H., Li, Y., Shi, J. (2006). Copper (II) and lead (II) removal from aqueous solution in fixed-bed columns by manganese oxide coated zeolite, *J. Hazard. Mater.*, **137**, pp. 934–942.

Hritcu, D., Humelnicu, D., Dodi, G., Popa, M.L. (2012). Magnetic chitosan composite particles: Evaluation of thorium and uranyl ion, *Carbohyd. Polym.*, **87**, pp. 1185–1191.

Hu, J., Chen, G., Lo, I.M.C. (2005). Removal and recovery of Cr(VI) from wastewater by maghemite nanoparticles, *Water Res.*, **39**, pp. 4528–4536.

Humelnicu, D., Dinu, M.V., Dragan, E.S. (2011). Adsorption characteristics of UO_2^{2+} and Th^{4+} ions from simulated radioactive solutions onto chitosan/clinoptilolite sorbents, *J. Hazard. Mater.*, **185**, pp. 447–455.

Hyung, H., Fortnter, J.D., Hughes, J.B., Kim, J.H. (2007). Natural organic matter stabilizes carbon nanotubes in the aqueous phase, *Environ. Sci. Technol.*, **41**, pp. 179–184.

Ilyina, A.V., Tikhonov, V.E., Varlamov, V.P., Radigina, L.A., Tatarinova, N.Y., Yamskov, I.A. (1995). Preparation of affinity sorbents and isolation of individual chitinases from a crude supernatant produced by *Streptomyces kurssanovii* by a one-step affinity chromatographic system, *Biotechnol. Appl. Biochem.*, **21**, pp. 139–148.

Jeon, C., Holl, W.H. (2003). Chemical modification of chitosan and equilibrium study for mercury ion removal, *Water Res.*, **37**, pp. 4770–4780.

Jeon, C., Holl, W.H. (2004). Application of the surface complexation model to heavy metal sorption equlibria onto aminated chitosan. *Hydrometallurgy*, **71**, pp. 421–428.

Kongsrichroen, N., Polprasrert, C. (1996). Chromium removal by a bipolar electrochemical precipitation process, *Water Sci. Technol.*, **34**, pp. 109–116.

Kortenkamp, A., Casadevall, M., Faux, S.P., Jennar, A., Shayer, R.O.J., Woodbridge, N., Obrien, P.A. (1996). Role for molecular oxygen in the formation of DNA damage during the reduction carcinogen of the chromium (VI) by glutathione, *Arch. Biochem. Biophys.*, **329**, pp. 199–207.

Krishnapriya, K.R., Kandaswamy, M. (2009). Synthesis and characterization of a cross-linked chitosan derivative with a complexing agent and its adsorption studies toward metal (II) ions, *Carbohyd. Res.*, **344**, pp. 1632–1638.

Kuchta, B., Firlej, L., Maurin, G. (2005). Modeling of adsorption in nanopores, *J. Mol. Model*, **11**, pp. 293–300.

Kushwaha, S., Sudhakar, P.P. (2011). Adsorption of mercury(II), methyl mercury(II) and phenyl mercury(II) on chitosan cross-linked with a barbital derivative, *Carbohyd. Polym.,* **86**, pp. 1055–1062.

Kwok, K.C.M., Lee, V.K.C., Gerente, C., Mckay, G. (2009). Novel model development for sorption of arsenate on chitosan, *Chem. Eng. J.,* **151**, pp.122–133.

Langmuir, I. (1918). The adsorption of gases on plane surfaces of glass, mica and platinum, *J. Am. Chem. Soc.,* **40**, p. 1361.

Liu, L., Li, C., Bao, C., Jia, C., Xiao, P., Liu, X., Zhang, Q. (2012). Preparation and characterization of chitosan/graphene oxide composites for the adsorption of Au(III) and Pd(II), *Talanta.,* **93**, pp. 350– 357.

Low, K.S., Lee, C.K., Leo, A.C. (1995). Removal of metals from electroplating wastes using banana pith, *Bioresour. Technol.,* **51**, pp. 227–231.

Marshall, W.E., Johns, M.M. (1996). Agricultural by-products as metal adsorbents: Sorption properties and resistance to mechanical abrasion, *J. Chem. Technol. Biotechnol.,* **66**, pp. 192–198.

Metwally, E., Elkholy, S.S., Salem, H.A.M., Elsabee, M.Z. (2009). Sorption behavior of ^{60}Co and $^{152+154}$Eu radionuclides onto chitosan derivatives, *Carbohyd. Polym.,* **76**, pp. 622–631.

Milosavljević, N.B., Ristić, M.D., Perić-Grujić, A.A., Filipović, J.M., Štrbac, S.B., Rakočević, Z.L., Krusic, M.T.K. (2011). Removal of Cu^{2+} ions using hydrogels of chitosan, itaconic and methacrylic acid: FTIR, SEM/EDX, AFM, kinetic and equilibrium study, *Colloids Surf. A. Physcicochem. Eng. Aspects,* **388**, pp. 59–69.

Modrzejewska, Z., Kaminski, W. (1999). Separation of Cr(VI) on chitosan membranes, *Ind. Eng. Chem. Res.,* **38**, pp. 4946–4950.

Mohan, D., Singh, K.P., Singh, V.K. (2005). Removal of hexavalent chromium from aqueous solution using low cost activated carbons derived from agricultural waste materials and activated carbon fabric cloth, *Ind. Eng. Chem. Res.,* **44**, pp. 1027–1042.

Monser, L., Adhoum, N. (2002). Modified activated carbon for the removal of copper, zinc, chromium and cyanide from wastewater, *Sep. Purif. Technol.,* **26**, pp. 137–146.

Muzzarelli, R.A.A. (2011). Potential of chitin/chitosan-bearing materials for uranium recovery: An interdisciplinary review, *Carbohyd. Polym.,* **84**, pp. 54–63.

Namdeo, M., Bajpai, S.K. (2008). Chitosan-magnetite nanocomposites (CMNs) as magnetic carrier particles for removal of Fe(III) from aqueous solutions, *Colloids Surf. A. Physcicochem. Eng. Aspects,* **320**, pp. 161–168.

Nomanbhay, S.M., Palanisamy, K. (2005). Removal of heavy metal from industrial wastewater using chitosan-coated oil palm shell charcoal, *Electron. J. Biotechnol.*, **8**, pp. 44–53.

Pandey, S., Mishra. S.B. (2011). Organic-inorganic hybrid of chitosan/organoclay bionanocomposites for hexavalent chromium uptake, *J. Colloid. Interface Sci.*, **361**, pp. 509–520.

Parsons, J.G., Hejazi, M., Tiemann, K.J., Henning, J., Gardea-Torresdey, J.L. (2002). An XAS study of the binding of copper(II), zinc(II), chromium(III) and chromium(VI) to hops biomass, *Microchem. J.*, **71**, pp. 211–219.

Paulino, A.T., Santos, L.B., Nazaki, J. (2008). Removal of Pb^{2+}, Cu^{2+}, and Fe^{3+} from battery manufacture wastewater by chitosan produced from silkworm chrysalides as a low-cost adsorbent, *React. Funct. Polym.*, **68**, pp. 634–642.

Popuri, S.R., Vijaya, Y., Boddu, V.M., Abburi, K. (2009). Adsorptive removal of copper and nickel ions from water using chitosan-coated PVC beads, *Bioresour. Technol.*, **100**, pp. 194–199.

Prakash, S., Chakrabarty, T., Singh, A.K., Sasi, V.K. (2012). Silver nanoparticles built-in chitosan modified glassy carbon electrode for anodic stripping analysis of As(III) and its removal from water, *Electrochim. Acta.*, **72**, pp. 157–164.

Qu, R.J., Sun, C.M., Wang, M.H., Ji, C.N., Xu, Q., Zhang, Y., Wang, C., Chen, H., Yin, P. (2009). Adsorption of Au(III) from aqueous solution using cotton fiber/chitosan composite adsorbents, *Hydrometallurgy*, **100**, pp. 65–71.

Rajiv Gandhi, M., Kousalya, G.N., Meenakshi, S. (2011). Removal of copper (II) using chitin/chitosan nano-hydroxyapatite composite, *Int. J. Biol. Macromol.*, **48**, pp. 119–124.

Rao, M., Parwate, A.V., Bhole, A.G. (2002). Removal of Cr^{6+} and Ni^{2+} from aqueous solution using bagasse and fly ash, *Waste Manage.*, **22**, pp. 820–821.

Rengaraj, S., Joo, C.K., Kim, Y., Yi. J. (2003). Kinetics of removal of chromium from water and electronic process wastewater by ion exchange resins: 1200H, 1500H and IRN97H, *J. Hazard. Mater.*, **102**, pp. 257–287.

Rengaraj, S., Yeon, K.H., Moon, S.H. (2008). Removal of chromium from water and wastewater by ion exchange resins, *J. Hazard. Mater.*, **87**, pp. 273–287.

Salehi, E., Madaeni, S.S., Rajabi, L., Vatanpour, V., Derakhshan, A.A., Zinadini, S., Ghorabi, S.H., Ahmadi Monfared, H. (2012). Novel chitosan/poly (vinyl) alcohol thin adsorptive membranes modified with amino functionalized multi-walled carbon nanotubes for Cu(II) removal from water: Preparation, characterization, adsorption kinetics and thermodynamics, *Sep. Purif. Technol.* **89**, pp. 309–319.

Salem, M.A., Makki, M.S.I., Abdelaal, M.Y.A. (2011). Preparation and characterization of multi-walled carbon nanotubes/chitosan nanocomposite and its application for the removal of heavy metals from aqueous solution, *J. Alloys Compd.*, **509**, pp. 2582–2587.

Sankararamakrishnan, N., Dixit, A., Iyengar, L., Sanghi, R. (2006). Removal of hexavalent chromium using novel crosslinked Xanthated chitosan, *Bioresour. Technol.*, **97**, pp. 2377–2382.

Sawalha, M.F., Gardea-Torresdey, J.L., Parsons, J.G., Saupe, G., Peralta-Videa, J. R. (2005). Determination of adsorption and speciation of chromium species by saltbush (*Atriplex canescens*) biomass using a combination of XAS and ICP OES. *Microchem. J.*, **81**, pp. 122–132.

Septhum, C., Rattanaphani, S., Bremner, J.B., Rattanaphani, V. (2007). An adsorption study of Al(III) ions onto chitosan, *J. Hazard. Mater.*, **148**, pp. 185–191.

Sharma, Y.C., Singh, B., Agarwal, A., Weng, C.H. (2008). Removal of chromium by riverbed sand and wastewater effect of important parameters, *J. Hazard. Mater.*, **151**, pp. 789–793.

Spinelli, V.A., Laranjeira, M.C.M., Favere, V.T. (2004). Preparation and characterization of quaternary chitosan salt: Adsorption equilibrium of chromium (VI) ion, *React. Funct. Polym.*, **61**, pp. 347–352.

Sugunan, A., Thanachayanont, C., Dutta, J., Hilborn, J.G. (2005). Heavy-metal ion sensors using chitosan-capped gold nanoparticles, *Sci. Technol. Adv. Mater.*, **6**, pp. 335–340.

Sureshkumar, M.K., Das, D., Mallia, M.B., Gupta, P.C. (2010). Adsorption of uranium from aqueous solution using chitosan-tripolyphosphate (CTPP) beads, *J. Hazard. Mater.*, **184**, pp. 65–72.

Tang, H., Chang, C., Zhang, L. (2011). Efficient adsorption of Hg^{2+} ions on chitin/cellulose composite membranes prepared via environmentally friendly pathway, *Chem. Eng. J.*, **173**, pp. 689–697.

Vieira, R.S., Oliveira, M.L.M., Guibal, E., Castellon, E.R., Beppu, M.M. (2011). Copper, mercury and chromium adsorption on natural and crosslinked chitosan films: An XPS investigation of mechanism, *Colloids Surf. A. Physicochem. Eng. Aspects*, **374**, pp. 108–114.

Volesky, B. (2001). Detoxification of metal-bearing effluents: Biosorption for the next century, *Hydrometallurgy*, **59**, pp. 203–216.

Wan, M-W., Kan, C-C., Rogel, B.D., Dalida, M.L.P. (2010). Adsorption of copper (II) and lead (II) ions from aqueous solution on chitosan-coated sand, *Carbohydr. Polym.,* **80**, pp. 891–899.

Wang, G., Liu, J., Wang, X., Xie, J., Deng, N. (2009). Adsorption of uranium (VI) from aqueous solution onto cross-linked chitosan, *J. Hazard. Mater.,* **168**, pp. 1053–1058.

Wang, J.S., Peng, R.T., Yang, J.H., Liu, Y.C., Hu, X.J. (2011). Preparation of ethylenediamine-modified magnetic chitosan complex for adsorption of uranyl ions, *Carbohyd. Polym.,* **84**, pp. 1169–1175.

Wen, Y., Tang, Z., Chen, Y., Gu, Y. (2011). Adsorption of Cr(VI) from aqueous solutions using chitosan-coated fly ash composite as biosorbent, *Chem. Eng. J.,* **175**, pp. 110–116.

Yan, H., Yang, L., Yang, G., Yang, H., Li, A., Cheng, R. (2012). Preparation of chitosan/poly(acrylic acid) magnetic composite microspheres and applications in the removal of copper (II) ions from aqueous solutions, *J. Hazard. Mater.,* **229-230**, pp. 371–380.

Yi, H., Wu, L., Bentley, W.E., Ghodssi, R., Rubloff, G.W., Culver, J.N., Payne, G.F. (2005). Biofabrication with chitosan. *Biomacromolecules,* **9**, pp. 2881–2894.

Yuwei, C., Jianlong, W. (2011). Preparation and characterization of magnetic chitosan nanoparticles and its application for Cu(II) removal, *Chem. Eng. J.,* **168**, pp. 286–292.

Zhang, D., Wei, S., Kaila, C., Su, X., Wu, J., Karki, A.B., Young, D.P., Guo, Z. (2010). Carbon-stabilized iron nanoparticles for environmental remediation, *Nanoscale,* **2**, pp. 917–919.

Zhang, L., Yang, S., Han, T., Zhong, L., Ma, C., Zhou, Y., Han, X. (2012). Improvement of Ag(I) adsorption onto chitosan/triethanolamine composite sorbent by an ion-imprinted technology, *Appl. Sur. Sci.,* **263**, pp. 696–703.

Zhao, Y., Peralta-Videa, J.R., Lopez-Moreno, M.L., Ren, M., Saupe, G., Gardea-Torresdey, J.L. (2011). Kinetin increases chromium absorption, modulates its distribution, and changes the activity of catalyse and ascorbate peroxidase in Mexican palo verde, *Environ. Sci. Technol.,* **45**, pp. 1082–1087.

Zheng, W.X.M., Li, Q., Yang, G.M., Zeng, X.X., Shen, J., Zhang, J., Liu, J. (2007). Adsorption of Cd(II) and Cu (II) from aqueous solution by carbonate hydroxyapatite derived from egg shell waste, *J. Hazard. Mater.,* **147**, pp. 534–539.

Zhou, L., Wang, Y., Liu, Z., Hu, Q. (2006). Carboximethylchitosan-Fe_3O_4 Chitosan nanoparticles; preparation and adsorption behaviour toward Zn ions, *Acta Phis. Chim. Sin.,* **22,** pp. 1342–1348.

Zimmermann, A.C., Mecabo, A., Fagundes, T., Rodrigues, C.A. (2010). Adsorption of Cr(VI) using Fe-crosslinked chitosan complex (Ch-Fe), *J. Hazard. Mater.,* **179**, pp. 192–196.

Zou, X., Pan, J., Ou, H., Wan, X., Guan, W., Lee, C., Yan, Y., Duan, Y. (2011). Adsorptive removal of Cr(III) and Fe(III) from aqueous solution by chitosan/attapulgite composites: Equilibrium, thermodynamics and kinetics, *Chem. Eng. J.,* **167**, pp. 112–121.

Chapter 2

Gum-Polysaccharide-Based Nanocomposites for the Treatment of Industrial Effluents

Hemant Mittal,[a,b] Balbir Singh Kaith,[a] Ajay Kumar Mishra,[b] and Shivani Bhardwaj Mishra[b]
[a]Department of Chemistry,
National Institute of Technology, Jalandhar 144011, India
[b]Department of Applied Chemistry,
University of Johannesburg, Doornfontein 2028, South Africa
bskaith@yahoo.co.in

The polymers that change their properties in response to external environmental conditions are termed smart polymers. Smart polymers are also known as stimuli-responsive or environment-sensitive polymers and have been used widely in the area of biotechnology, pharmaceuticals, biomedicals, and bioengineering. Polysaccharide-based cross-linked networks are one of the most important classes of nanocomposites and have got significant applications. Contamination of water by industrial effluents such as heavy metal ions, dyes, and polyaromatic molecules is one of the most serious environmental issues and has adverse effects on human health. Water pollution has become the major source of

---

Nanocomposites in Wastewater Treatment
Edited by Ajay Kumar Mishra
Copyright © 2015 Pan Stanford Publishing Pte. Ltd.
ISBN 978-981-4463-54-6 (Hardcover), 978-981-4463-55-3 (eBook)
www.panstanford.com

concern and a priority for most of the industries. For last few years, scientists from all over the world are working on the removal of toxic pollutants from the polluted water, and adsorption is one of the most preferred methods. Natural polysaccharides and their derivatives-based cross-linked nanocomposites are known to be one of the best absorbents and the method is a low cost green technology.

2.1 Introduction

Polysaccharide-based cross-linked nanocomposites are water insoluble at physiological pH and temperature but swell up to several times of their dry weight. Absorption of water is one of the most important properties of such nanocomposites. Because of the capacity to retain large content of biological fluids, they constitute a class of the most excellent biocompatible materials. Their water absorption capacity is due to the presence of certain hydrophilic groups such as –COOH, –NH$_2$, and –OH [1, 2]. The water absorption capacity also depends on the flexibility of the polymer chains, cross-linking density, and free volume between the polymer chains [3, 4]. It was observed that the extent of water absorption mainly depends on the extent of cross-linking between different polymeric chains, affinity of the polymer for the water molecules, concentration of the fixed charges in the dry polymer, and ionic strength of the solvent. The nature of the starting materials also plays an important role in the swelling of the nanocomposites [5]. One of the most important classes of polysaccharide-based nanocomposites is smart or responsive nanocomposites. Responsive nanocomposites undergo change in their volume in response to a very small change in environmental conditions such as pH, temperature, electric currents, and ionic strength. Responsive nanocomposites are widely used in many industrial applications such as drug delivery, soft tissue engineering, wastewater treatment, and many biomedical applications [6–8].

Nanocomposites can be synthesized from natural and synthetic sources. Synthetic–polymer-based nanocomposites have many advantages over natural polysaccharide- or biopolymer-based nanocomposites, such as better mechanical strength, high resistivity to chemicals, better stereospecific applications, and longer life. But synthetic nanocomposites suffer from some serious disadvantages such as nondegradability and cause environmental pollution, which is one of the major threats to the modern world especially in the developed countries. So biopolymer-based nanocomposites have been widely used as a replacement to synthetic polymers because of their low cost, nontoxicity, inherent biocompatibility, and biodegradability [9, 10]. Nowadays scientists are giving more emphasis on biodegradable polymers such as polysaccharides, which have been extensively used and investigated for industrial applications. Various natural polysaccharides such as guar gum, gum xanthan, chitosan, and alginate have been modified for their wide applications in a variety of areas such as wastewater treatment, biomedicine, food industry, pharmaceuticals, and biotechnological fields [11, 12]. This chapter includes the characteristics and structural properties of polysaccharides, various techniques of their modification and their utilization for the treatment of polluted water, which will force researchers to pay attention on the utilization of these biopolymers in different areas of research.

2.1   Gum Polysaccharides

Polysaccharides consist of different monosaccharide units that are joined together by O-glycosidic linkages. Diversity in the structural features of different polysaccharides is because of the difference in the monosaccharide composition, chain shapes, degree of polymerization, and linkage types. Polysaccharides are mainly used in food industries and pharmaceutics. Therefore, they are also known as food hydrocolloids. Polysaccharides are mainly obtained from plants as exudates or extracellular substances and from microorganisms as their secreted products. Polysaccharide-based biopolymers have a branched structure in which a linear chain of a monosaccharide unit is attached to the side chains of other different monosaccharide units at certain positions. Since the branched polymers occupy lesser volume as compared to linear chain polymers of almost same molecular weight, they form more viscous solution than the linear polymers of same molecular weight. Many of the polysaccharides swell in water to a large extent, so they are found suitable for the preparation of nanocomposites. Chemical modification of different polysaccharides through different techniques such as graft copolymerization provides hydrocolloids with improved functionality and stability [13, 14].

2.1.1 Gum Arabic

Gum arabic is obtained from *Acacia senegal* and *Acacia seyal* as a dry exudate. About 500 species of Acacia are found in Africa, India, Australia, Central America, and southwest North America. It is highly soluble in water and gives pale yellow to orange-brown clear solution. Highly branched complex structure of gum arabic is composed of D-galactopyranose (44%), L-arabino-pyranose, furanose (25%), L-rhamnopyranose (14%) and D-glucopyranosyl uronic acid (15.5%), and 4-O-methyl-D-glucopyranosyl uronic acid (1.5%) [15]. The primary structure of gum has the main chain of repeated (1→3)-β-D-glactopyranosyl unit with side chains of L-arabinofuranosyl, L-rhamnopyranosyl, D-glactopyranosyl, and D-glucopyranosyl uronic acid at the C-6 position (Fig. 2.1). Integral part of gum arabic also contains protein (about 2%). It is mainly used as emulsifier and stabilizer in beverage industry [16].

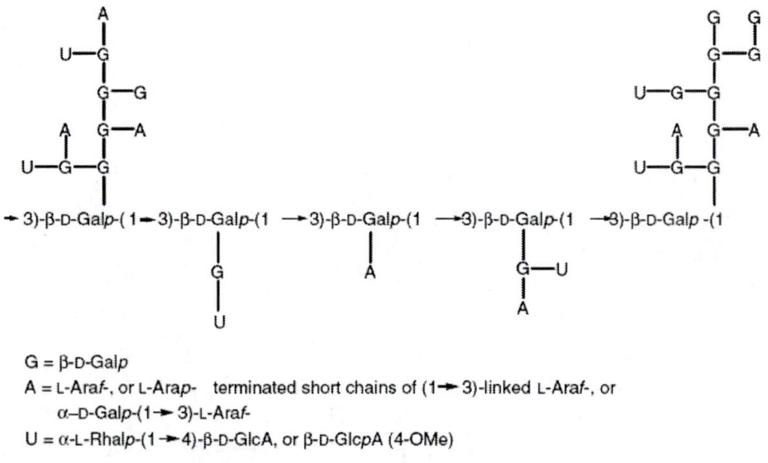
→3)-β-D-Galp-(1→3)-β-D-Galp-(1 →3)-β-D-Galp-(1 →3)-β-D-Galp-(1 →3)-β-D-Galp -(1
G = β-D-Galp
A = L-Araf-, or L-Arap- terminated short chains of (1→3)-linked L-Araf-, or α–D-Galp-(1→3)-L-Araf-
U = α-L-Rhalp-(1→4)-β-D-GlcA, or β-D-GlcpA (4-OMe)

Figure 2.1 Structure of gum arabic. From ref. [17].

2.1.2 Gum Karaya

Gum karaya is obtained from *Sterculia urens* of *Sterculiaceae* family. This plant is mainly found in the Indian subcontinent and North African countries. Solubility of gum karaya in water is very low, but it swells many times of its original weight to give dispersion [18, 19]. It is a highly branched complex polysaccharide consisting of D-galacturonic acid, D- galactose, L-rhamnose, and D-glucoronic acid. The carbohydrate structure has the main chain of rhamnogalacturonan consisting of α-(1→4)-linked D-galacturonic acid and α-(1→2)-linked L-rhamnosyl residues. The side chain is made of (1→3)-linked β-D-glucuronic acid, or (1→2)-linked β-D-galactose on the galacturonic acid unit where one-half of the rhamnose is substituted by (1→4)-linked β-D-galactose (Fig. 2.2) [20]. Gum karaya is mainly used as stabilizer and emulsifier in pharmaceutical industries, as a food additive, in manufacturing long-fibered light weight papers, and in the textile industry.

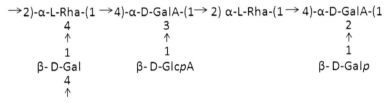

Figure 2.2 Structure of gum karaya. From ref. [20].

2.1.3 Gum Tragacanth

Gum tragacanth is obtained as a gummy exudate of *Astragalus gammifer* or other asiatic species of *Astragalus* genus. It is a highly branched heterocyclic polysaccharide [21]. It is found in dry dessert and mountainous regions of southwest Asia and mainly in Iran and Turkey [22]. *Astragalus* genus is a low bushy perennial small shrub and has a large tap root with branches. It consists of water soluble tragacanthin and water insoluble tragacanthic acid. Tragacanthin dissolves in water and forms a viscous solution, whereas tragacanthic acid swells to give a soft and adhesive gel like state [23]. Tragacanthin is a highly branched neutral polysaccharide consisting of the main chain of repeated D-galactose residues to which chains of L-arabinose residues are attached and have spheroidal molecular shape. However, tragacanthic acid consists of a chain of repeated units of (1→4)-linked D-galactopyranosyl with a side chain of substituted xylosyl linked at the C-3 position of the glacturonic acid residues (Fig. 2.3).

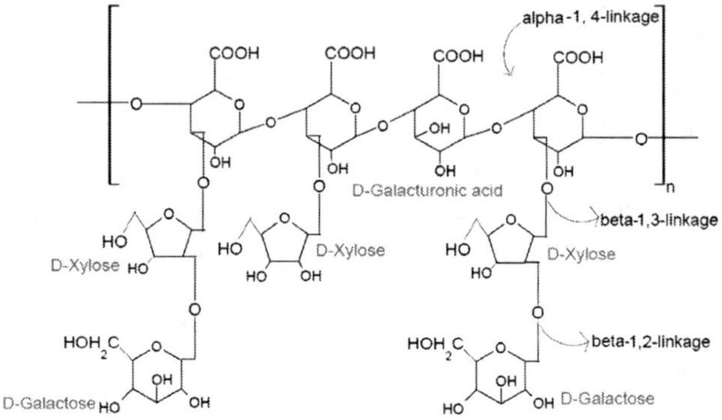

Figure 2.3 Structure of tragacanthic acid. From ref. [20].

2.1.4 Gum Xanthan

Gum xanthan is a high molecular weight animal polysaccharide, which is obtained from the microorganism *Xanthomous campestris* through fermentation process. It is highly soluble in water [24]. The structure of gum xanthan consists of the main chain of (1→4)-linked β-D-Glc*p* residues substituted at O-3 of alternate glucose residues with β-D-Man*p*-(1→4)-β-D-Glc*p*A-(1→2)-α-D-Man*p*-(1→) unit (Fig. 2.4). Noncarbohydrate substituents include the acetyl group at the O-6 position of the inner Man*p* residue and the pyruvate group at the O-4,6 position of the terminal Man*p* [24].

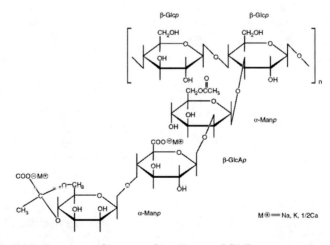

Figure 2.4 Structure of gum xanthan. From ref. [24].

The main chain in the structure of gum xanthan is same as that of cellulose, but it has a trisaccharide side chain on the alternate sugar unit. In this side chain, a glucoronic acid salt is attached between the mannose acetate and a terminal mannose unit. Gum xanthan is anionic because of the presence of glucoronic and pyruvic acids on the side chains. The solution of gum xanthan is found to be stable in both acidic and alkaline media between pH 2.0 and 12.0, whereas below pH 2.0 and above pH 12.0, the viscosity of gum xanthan solution decreases. Gum xanthan can be used with other gums to give elastic gels. But the use of gum xanthan with cellulose or its derivatives should be avoided because it contains many enzymes, which digest cellulose. Gum xanthan is mainly used in the food industry as a suspending and thickening agent [20]. Sand et al. [25] synthesized xanthan gum and 2-acrylamidoglycolin acid–based hydrogels using bromate/thiourea redox pair as an initiator. They also studied their flocculation behavior, metal ion absorption capacity, and resistance to biodegradation. In another study, Jindal et al. [26] synthesized gum xanthan and acrylamide-based hydrogels and studied their swelling behavior in response to change in pH and temperature. They also studied their swelling behavior in response to change in the ionic strength and cationic charge in different chloride salt solutions.

2.1.5 Gum Gellan

Gum gellan is obtained from the microorganism *Auromonas elodea* as an extracellular bacterial polysaccharide. It is present in substituted and unsubstituted form. It is produced by inoculating a fermentation medium with *Auromonas elodea*. Fermentation is carried out under controlled conditions, and the product is recovered in a few days. Recovery by alcohol precipitation yields the substituted gum, whereas recovery by alkali produces the unsubstituted gum. When hot solutions of gum gellan are cooled in the presence of gel promoting cations, it forms a gel. Substituted form produces soft, elastic gels, whereas unsubstituted form produces hard, brittle gels, and the gel properties depend on the degree of substitution. The primary structure of gum gellan is composed of a repeating sequence of →3)-β-D-Glcp-(1→4)-β-D-GlcpA-(1→4)-β-D-Glcp-(1→4)-α-L-Rhap-(1→ units (Fig. 2.5) [27]. Gum gellan is used as a gelling agent in dessert jellies, dairy products, and sugar confectionery. It is also used to prepare structured lipids [28].

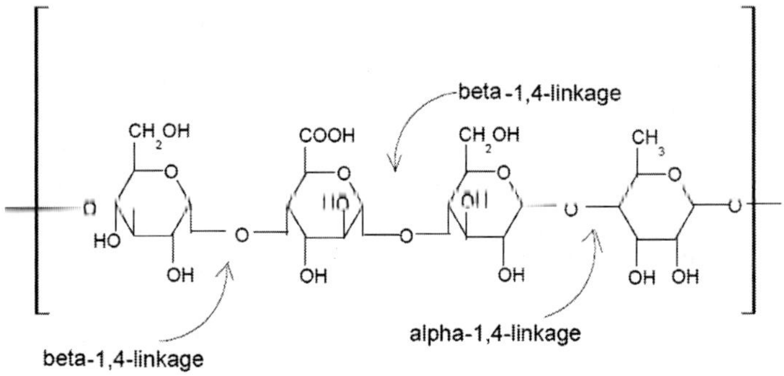

Figure 2.5 Structure of gum gellan. From ref. [27].

2.1.6 Guar gum

Guar gum is extracted from the seed of the plant *Cyamopsis tetragonoloba*. It is mainly found in the sandy soils of North India and some parts of Pakistan. It is a nonionic hydrocolloid and is not affected by ionic strength but gets degraded at pH 3.0 and 50 °C [29]. Guar gum is a galactomannan, and its chemical structure is composed of a (1→4)-linked β-D-mannopyranose backbone with branches of (1→6)-linked α-D-galactopyranose (Fig. 2.6). There are between 1.5 and 2 mannose residues for every galactose residue. A lot of work has been reported on the graft copolymerization of guar gum and their utilization in different industries [30].

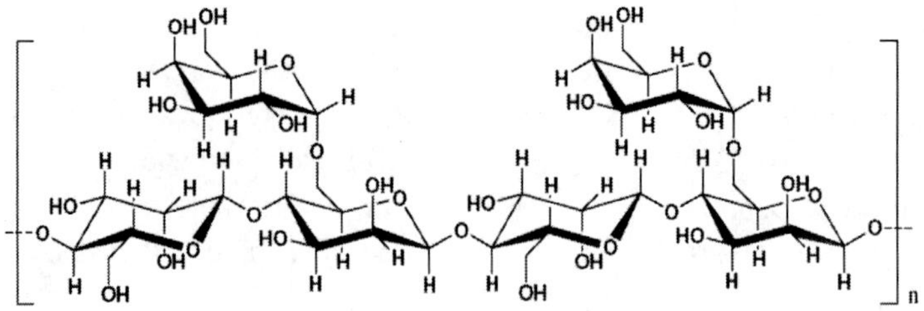

Figure 2.6 Structure of guar gum. For ref. [33].

Chauhan et al. [31] synthesized guar gum–based hydrogels and utilized them for the uptake of Cu^{2+} ions in response to change in pH, temperature, and ionic strength. The maximum sorption capacity of 125.893 mg/g was observed. The graft copolymer of guar gum with poly(methyl methacrylate) was synthesized using persulphate/ascorbic acid redox pair as an initiator and utilized it for the removal of Cr(VI) ions from aqueous solution [32].

2.1.7 Locust bean gum

Locust bean gum is obtained from the seeds of the carob tree of the *Ceretonia siliqua* family as an exudate. It is found in the Mediterranean regions. Only a little proportion of locust bean gum is soluble in cold water; therefore, it forms a highly viscous solution with water. The main chain of locust bean gum consists of (1→4)-linked β-D-mannose residues with the side chain of (1→6)-linked α-D-galactose (Fig. 2.7). The side chains of galactose sugars are clustered in blocks and are not distributed uniformly throughout the polymer chains, so the structure of chains is irregular with alternating "smooth" and substituted zones [34]. Locust bean gum is used in many food products, cosmetics, and pet foods. Because of the poor solubility of locust bean gum in water, very little work has been done on the graft copolymerization of locust bean gum. Water absorption of locust bean gum was studied by Torres et al. [35]. The edible films of kappa-carrageenan and locust bean gum were synthesized so as to have the properties of both the polysaccharides. The synthesized biopolymer was characterized using FTIR, DMA, and TGA techniques [36].

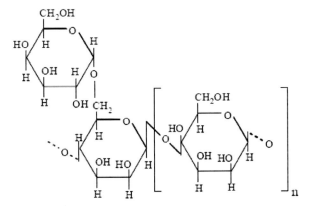

Figure 2.7 Structure of locust bean gum. From ref. [34].

2.1.7 Gum Ghatti

Gum ghatti, also known as Indian gum, is an exudate of the *Anogeissus latifolia* tree belonging to the family *Combretaceae*. It occurs as a gray to redish-gray powder, granular or light- to dark-brown lump and is almost odorless. Aspinall et al. [37] extensively studied the structure of gum ghatti and found it to have an extremely complex structure composed of sugars such as L-arabinose, D-galactose, D mannose, D-xylose, and D-glucoronic acid in a 48:29:10:5:10 molar ratio. This gum contains alternating 4-O-substituted and 2-O-subsituted α-D-mannopyranose units and chains of (1→6)-linked β-D-galactopyranose units with side chains, which are most frequently single L-arabinofuranose residues (Fig. 2.8) [38].

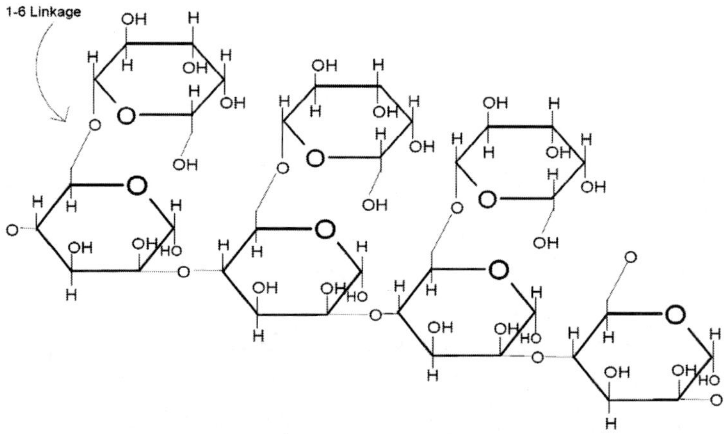

Figure 2.8 Structure of gum ghatti. From ref. [20].

Anogeissus latifolia is mainly found in all regions of India and Sri Lanka and has the largest forest coverage in India. It is one of the most useful trees in India. Its leaves contain a large amount of tannins and are used in India for tanning. Gum ghatti is one of the oldest and best known natural gum substances. The availability and harvest depend directly on the monsoon season. While the plant needs a hardy amount of rain to grow, too much water will flood it out. Gum ghatti is exuded naturally in the form of rounded tears of less than 1 cm diameter or in the form of larger vermiform masses. It has a glassy fracture, and the color of the exudates varies from very light to dark brown. It is collected manually in the same way as most of other tree exudate gums. Gum ghatti was originally developed around 1900 as a substitute for gum arabic. However, due to the variation in the solubility and viscosity, it has never been accepted as major gum tree. The viscosity of grade-I gum ghatti was shown to vary from 30 cP to 400 cP for 5% solution [39]. Nodules of gum ghatti are not entirely soluble in cold water, but breaking up and heating at 90 °C improves its solubility. Kaith et al. [40] synthesized gum ghatti and acrylamide-based hydrogels and studied their swelling behavior in response to change in pH and temperature. They utilized them for the selective removal of saline from different petroleum fraction-saline emulsions.

2.2 Stimuli-Responsive Nanocomposites

Nanocomposites that abruptly change their physical or chemical properties in response to the small changes in the external environmental stimulus such as pH, temperature, electric field, and magnetic field are known as stimuli-responsive nanocomposites [41, 42]. They are also known as smart, environmentally responsive, stimuli-sensitive, and intelligent nanocomposites. Stimuli-responsive nanocomposites have a wide range of industrial applications such as drug delivery, wastewater treatment, and chromatography [43, 44]. The hydrogel initially recognizes the signal, then judges the magnitude of that signal, and finally changes its chain conformation in response to that signal. The external stimuli could be chemical or physical. Chemical stimuli such as pH, chemical agents, or ionic strength of the swelling medium can change the interactions between the polymer chain and solvent molecules. However, physical stimuli such as temperature, electric, or magnetic fields on alteration affect the molecular interactions, thereby causing change in molecular shape.

2.2.1 Temperature-Responsive Nanocomposites

Temperature-responsive nanocomposites are one of the most important and most widely studied classes of stimuli-responsive nanocomposites because it is easy to control a change in temperature in vitro and in vivo [45]. The most important characteristic of the temperature-responsive nanocomposites is the intermolecular interaction between the three-dimensional polymer network and swelling medium, which ultimately results in the physical cross-links, micelle aggregation, or polymer shrinkage. Basically, there are two types of intermolecular forces, namely, hydrogen bonding and hydrophobic interactions, which take place between water and nanocomposites. Hydrogen bonding by intermolecular association is a random coil-to-helix transition in which different biopolymer chains form a helix conformation and results in the physical cross-links to make a polymer network by lowering the temperature [46]. Also association/dissociation of the hydrogen bonding between different pendant groups can be controlled by change in temperature. Moreover, hydrophilic groups of the hydrogels form hydrogen bonds with water molecules. Such bonds cooperatively form a stable shell of hydration around the hydrophilic groups leading to the greater uptake of water and result in larger swelling. However, the associated interactions among the hydrophilic groups release the entrapped water molecules from the nanocomposites network by change in temperature [47]. Using this mechanism, Chen and Cheng [48] studied the swelling/de-swelling behavior of the thermo-responsive nanocomposites of chitosan and hyaluronic acid. Critical solution temperature is one of the most important properties of the temperature-responsive nanocomposites. Critical solution temperature is defined as the temperature at which different phases of the polymer and solution are separated according to their composition. The temperature above which the polymer goes a phase transition from a soluble state to insoluble is called lower critical solution temperature (LCST). Poly(N-isopropylacrylamide) (PNIPAM) has LCST at 32 °C as it exhibits a sharp phase transition in water at about 32 °C [49]. Yue et al. [50] synthesized thermo-sensitive nanocomposites composed of poly(N-isopropylacrylamide-co-acrylic acid) (PVA) microsphere matrices and embedded PNIPAM nanogels, which have different volume phase transition temperatures, and the volume phase transition temperature of PNA microsphere was found to be lower than that of PNIPAM nanogels; therefore, the nanogels shrink earlier than that of microshperes. Chuang et al. [51] synthesized chitosan-based thermo-responsive nanocomposites and utilized them for the release of doxycycline hyclate.

2.2.2 pH-Responsive Nanocomposites

The pH-responsive nanocomposites are a class of responsive nanocomposites whose swelling properties change in response to a small change in pH. They consist of ionizable functional groups, which can accept or donate protons upon changing the pH of the swelling medium. With change in the environmental pH, the pK_a value of the weakly ionizable functional group in the nanocomposite network changes and results in the rapid swelling/de-swelling of the respective nanocomposite systems with the alteration of the hydrodynamic volume of the polymer chains [52]. This change in the volume of the nanocomposites is explained on the basis of osmotic pressure swelling (π_{ion}) theory resulting from difference between the concentration of mobile ions in swollen polymer and the external solution [53]. The polymer backbones with ionizable groups form polyelectrolytes in water. Weak polyacids such as poly(acrylic acid) act as the proton acceptor in acidic medium, whereas act as the proton donor in neutral as well as in basic media. On the other hand, polybase groups act as a proton donor in acidic medium and as a proton acceptor in basic as well as in neutral media. Therefore, the selection of polyacids and polybases is very important for the desired applications.

In case of a backbone with polyacids, increased swelling was observed in neutral medium as compared to alkaline and acidic media. In dilute electrolyte solutions, swelling behavior of ionic polymers is due to the osmotic swelling pressure between the nanocomposite phase and external solution (swelling medium). The osmotic swelling pressure (π_{ion}) is given as [54]:

$$\pi_{ion} = RT\sum(C_i^g - C_i^s),$$

where C_i^g and C_i^s are the molar concentrations of mobile ions in the swollen nanoconposite and external solution, respectively. R is the gas constant, and T is the absolute temperature.

In neutral medium, π_{ion} becomes very large, because the concentration of mobile ions in the external solution (C_i^s) becomes almost negligible. Also, electrostatic repulsion between carboxylate ions promotes the swelling, so larger swelling is observed. However, in acidic medium, most of the polyacids get protonated; therefore, the concentration of mobile ions within the swollen nanocomposite

(C_i^g) is less as compared to the concentration of the mobile ions in the solution resulting in the smaller π_{ion} value. However, in alkaline solutions, polyacid groups completely dissociate, but at the same time, the concentration of Na⁺ and OH⁻ ions is also very high and leads to a decrease in π_{ion} and percent swelling. Moreover, in alkaline medium, a screening effect of the counter ions (Na⁺) shields the electrostatic repulsion between the carboxylate anions. As a result, a remarkable decrease in P_s is observed [55].

2.3 Preparation of Nanocomposites

Polysaccharide-based nanocomposites can be synthesized by a number of methods depending on the requirement of the candidate polymer.

2.3.1 Graft Copolymerization/Cross-Linking

Mostly "grafting onto" and "grafting from" methods have been used for the grafting of synthetic polymers on various biopolymers. The graft copolymerization method includes free radical polymerization and controlled/living radical polymerization. In this method, multifunctional cross-linking agents are mixed with ionic or neutral monomers. The polymerization reaction is initiated using redox initiator system, or low and high energy radiations are used [56]. The solvent acts as a heat sink and minimizes temperature control problems. Unreacted monomers, cross-linkers, and initiators are removed by successive washings with the suitable solvent. The graft copolymerization method is widely used for the preparation of a great variety of nanocomposites. Smart nanocomposites can be prepared according to the requirement by incorporating suitable monomers and backbone polymers. For example, temperature- and pH-responsive nanocomposites can be prepared by the graft copolymerization of cellulose with PDMAEMA in DMF, and the synthesized nanocomposites show temperature- and pH-responsive properties due to the presence of PDMAEMA [57]. Pal et al. [58] synthesized the nanocomposites based on nanosilica and the xanthan gum grafted with polyacrylamide and utilized them for the successful absorption of Pb²⁺ ions.

2.3.2 Suspension Polymerization

In this type of polymerization technique, the initiator is completely soluble in monomer, whereas both of them are not soluble in the reaction medium. In suspension polymerization, the monomer phase is suspended in the medium in the form of small droplets by using stirrer and droplet stabilizer. The ratio of volume of monomer phase to the polymerization medium is 0.1–0.5. The polymerization is then initiated at the desired temperature (200–1000 °C), and at the end of the reaction, the monomer droplets are converted directly into micro beads of size 1 μ–1 mm. The kinetics of suspension polymerization depends on the concentration of monomer diluents in the monomer droplets as in case of bulk or solution polymerization [59]. Examples of industrially important polymers produced by suspension polymerization include polystyrene, polyvinylchloride, polyacrylates, and poly(vinyl acetate) [60].

2.3.3 Polymer Coacervation Process

Polymer coacervation is the process in which a polymer solution of liquid–liquid phase separates into the coacervate phase (polymer-rich phase) and the equilibrium phase (polymer-lean phase). In this process, microscopic droplets of the coacervate phase are formed in the liquid phase, and the resulted colloidal dispersion is very unstable and has a strong tendency for coalescence. The size of the droplet forms can be controlled by physical gelation or chemical cross-linking. The polymer coacervation process is generally divided into two categories: simple coacervation and complex coacervation.

2.3.3.1 Simple coacervation process

Simple coacervation is also known as segregative phase separation. In this method, the polymer is partially dissolved in the solution. This can be achieved by changing the solvent quality, solution temperature, or pH of the solvent in case of weak polyelectrolytes. This technique is frequently used in pharmaceutical technology to entrap drugs into microcapsules. The polymers with the low solvation degree in the segregated phase are regarded as precipitated polymer colloids and the polymers with LCST and UCST behavior can be precipitated by changing the temperature and pH of the solution. Sugihara et al. [61] photocross-linked the copolymers of poly(4-hydroxybutyl vinyl ether) and above the LCST, very fine coacervate droplets were formed. The coacervate droplets formed were cross-linked to obtain stable thermo-responsive nanocomposites.

2.3.3.2 Complex coacervation process

Complex coacervation process involves the associative phase separation of the two oppositely charged polyelectrolytes forming the polyelectrolyte complexes or di- and trivalent counter ions between polyelectrolytes. In some cases, complex coacervation occurs through the formation of hydrogen bonds between the polymers. By complex coacervation process, micro- and nanocomposites can be prepared without using surfactants and organic solvents under mild conditions. Du et al. [62] synthesized carboxymethyl konjac glucomannan–chitosan-based nanocomposites under very mild conditions using polyelectrolyte complexation. The prepared nanocomposites were found to be stable in water and successfully utilized as a drug delivery device. In another study, dextran sulfate and chitosan-based nanocomposites were prepared in aqueous solution. The prepared nanocomposite was utilized for the controlled delivery of insulin [63].

2.4 Utilization of Nanocomposites for Wastewater Treatment

The chemical contamination of water from a wide range of toxic substances such as heavy metals, organic compounds, and dyes is a serious problem because of their human toxicity [64]. Therefore, scientists from all over the world are developing techniques that can remove toxic pollutants from wastewater. Absorption of these heavy metal ions from industrial effluents by nanocomposites has been a great achievement. Adsorption by polysaccharide-based nanocomposites is proved to be a low cost technology for the purification of contaminated water, for extraction and separation of compounds, and a useful tool for protecting the environment [65].

The presence of excess amount of heavy metal ions such as Cu(II), Cr(VI), and Co(II) in surface and groundwater systems causes multiple health problems to people and aquatic life. Hg(II) and Cd(II) are considered most toxic heavy metal ions and damage the environment. Similarly, Cr(VI) is very toxic and is extremely harmful to bacteria, animals, plants, and human beings. Heavy metal ions are nonbiodegradable and tend to accumulate in living organisms thereby causing various diseases and disorders. Natural polysaccharide-based stimuli-responsive nanocomposites have been utilized by many research groups for the selective adsorption and removal of toxic heavy metals from industrial effluents, wastewater, and sea water [66, 67]. Polysaccharide-based nanocomposites are replacing other conventional methods because of their properties such as biodegradation, low cost, and easy availability. The metal ion uptake capacity of superabsorbent polymers mainly depends on the hydrophobic–hydrophilic balance and nature of monomers [68]. Different functionalities in the backbone polymers are generally introduced through graft copolymerization, and it helps the polymer to retain the metal ion by polymer analogue reaction or by the chelation absorption in the pores created within three-dimensional cross-linked networks. The selectivity in the absorption of a particular metal ion and its amount depends on the functionality introduced in the polymer through graft copolymerization. The graft copolymer of partially carboxymethylated guar gum with 2-acrylamidoglycolic acid was utilized for the selective adsorption of Pb^{2+}, Ni^{2+}, and Zn^{2+} ions, and the total uptake of metal ion was calculated by titrating the unabsorbed metal ions [69]. In another study, Sand et al. [25] used 2-acrylamidoglycolic acid and xanthan gum–based hydrogels for the absorption of Zn^{2+} ions. Starch and polyaniline-based nanocomposite was prepared by chemical oxidative polymerization for the removal of dyes such as black-5 and violet-4 from aqueous solution. The nanocomposite was found to remove 99% of black-5 and 98% of violet-4 from the dyes solution. The kinetics of the adsorption was found to follow the pseudo second-order model. The adsorption behavior was found to follow the modified Freundlich adsorption isotherm [70].

2.5 Conclusion

Polysaccharides are abundant in nature. They have many advantages such as biodegradability, biocompatibility, nontoxicity, and low cost over synthetic polymers. They are compatible to human body as they are easily degraded to monosaccharides by colonic bacteria. The environmental pollution caused by synthetic polymers is one of the biggest problems. The demand for biodegradable and environment friendly polysaccharide-based polymers is increasing day by day. Polysaccharides are associated with drawbacks such as moisture absorbance, minimum chemical resistance, low mechanical strength, and difficult processing, but their functionalization with vinyl monomers overcomes such limitations and can be utilized in various technological processes. Because of fluid absorption capacity and structural stability, smart nanocomposites have been found very efficient for the removal of toxic heavy metal ions from wastewater. The unique properties and structural features of nanocomposites make them suitable as potential device for diversified industrial applications.

References

1. Buchholz, F. L., and Graham, A. T. (1998). *Modern Superabsorbent Polymer Technology*, Wiley, New York.
2. Pourjavadi, A., and Kurdtabar, M. (2007). Collagen-based highly porous hydrogel without any porogen: Synthesis and characteristics, *Euro. Polym. J.*, **43**, pp. 877–889.
3. Tang, Q., Sun, X., Li, Q., Lin, J., and Wu, J. (2009). Preparation of porous polyacrylate/poly(ethylene glycol) interpenetrating network hydrogel and simplification of Flory theory, *J. Mater. Sci.*, **44**, pp. 3712–3718.
4. Chen, J., and Park, K. (2000). Synthesis and characterization of superporous hydrogel composites, *J. Control. Rel.*, **65**, pp. 73–82.
5. Flory, P. J. (1953). *Principles of Polymer Chemistry*, Cornel University Press, New York.
6. Li, T., Shi, X. W., Du, Y. M., and Tang, Y. F. (2007). Quaternized chitosan/alginate nanoparticles for protein delivery, *J. Biomed. Mater. Res. Part A*, **83A**, pp. 383–390.
7. Asoh, T., Kaneko, T., and Matsusaki, M. (2006). Rapid de-swelling of semi-IPNs with nanosized tracts in response to pH and temperature, *J. Control. Rel.*, **110**, pp. 387–394.
8. Durme, K. V., Mele, B. V., Loos, W., and Du Prez, F. E. (2005). Introduction of silica into thermo-responsive poly(N-isopropyl acrylamide)

hydrogels: A novel approach to improve response rates, *Polymer*, **46**, pp. 9851–9862.

9. Lee, W. F., and Chen, Y. J. (2001). Studies on preparation and swelling properties of the N-isopropylacrylamide/chitosan semi-IPN and IPN hydrogels, *J. Appl. Polym. Sci.*, **82**, pp. 2487–2496.

10. Mi, F. L., Tan, Y. C., Liang, H. F., and Sung, H. W. (2002). In vivo biocompatibility and degradability of a novel injectable-chitosan-based implant, *Biomater.*, **23**, pp. 181–191.

11. Krishnaiah, Y. S. R., and Srinivas, B. P. (2008). Effect of 5-fluorouracil pretreatment on the in vitro drug release from colon-targeted guar gum matrix tablets, *The Open Drug Del. J.*, **2**, pp. 71–76.

12. Kaith, B. S., Jindal, R., Mittal, H., and Kumar, K. (2012). Synthesis of cross-linked networks of *Gum ghatti* with different vinyl monomer mixtures and effect of ionic strength of various cations on its swelling behavior, *Int. J. Polym. Mater.*, **61**, pp. 99–115.

13. Mishra, M. M., Sand, A., Mishra, D. K., Yadav, M., and Behari, K. (2010). Free radical graft copolymerization of N-vinyl-2-pyrrolidone onto k-carrageenan in aqueous media and applications, *Carbohydr. Polym.*, **82**, pp. 424–431.

14. Jiangyang, F., Wang, K., Liu, M., and He, Z. (2008). In vitro evaluations of konjac glucomannan and xanthan gum mixture as the sustained release material of matrix tablet, *Carbohydr. Polym.*, **73**, pp. 241–247.

15. Whistler, R. L., and Bemiller, J. N. (1973). *Industrial Gums: Polysaccharides and Their Derivatives*, Academic Press, New York.

16. Fenyo, J. C., and Vandervelde, M. C. (1990). *Gums and Stabilizers for the Food Industry*, 5th edn., IRL Press, Oxford.

17. Cui, S. W. (2005). *Food Carbohydrates, Chemistry, Physical Properties and Applications*, Taylor and Francis, New York.

18. Anderson, D. M. W., and Wang, W. (1994). The tree exudate gums permitted in foodstuffs as emulsifiers, stabilisers and thickeners, *Chem. Ind. Forest Prod.*, **14**, pp. 73–84.

19. Le Cerf, D., Irinei, D., and Muller, G. (1990). Solution properties of gum exudates from *Sterculia urens* (karaya gum), *Carbohydr. Polym.*, **13**, pp. 375–386.

20. Weiping, W., and Branwell, B. (2000). Tragacanth and Karaya, in *Handbook of Hydrocolloids* (Phillips, G. O., and Williams, P. O., eds.), CRC Press, New York.

21. Whistler, R. L. (1993). Exudate Gums, in *Industrial Gums* (Whistler, R. L., and BeMiller, J. N., eds), Academic Press Incorporation, USA, pp. 309–340.

22. Anderson, D. M. W., and Bridgeman, M. M. E. (1985). The composition of the proteinaceous polysaccharides exuded by *Astragalus microcephalus*, *A. gummifer* and *A. kurdicus:* The sources of Turkish gum tragacanth, *Phytochem.*, **24**, pp. 2301–2304.

23. Aspinall, G. O., and Baillies, J. (1963). Gum tragacanth, Part I. Fraction of the gum and the structure of tragacanthic acid, *J. Chem. Soc.*, pp. 1702–1714.

24. Hui, Y. H., and Khachatourians, H. (1995). *Food Technology Microorganisms*, John Wiley–VCH Incorporation, USA.

25. Sand, A., Yadav, M., and Behari, K. (2010). Graft copolymerization of 2-acrylamidoglycolic acid onto xanthan gum and study of its physicochemical properties, *Carbohydr. Polym.*, **81**, pp. 626–632.

26. Jindal, R., Kaith, B.S., and Mittal, H. (2011). Rapid synthesis of acrylamide ontoxanthan gum-based hydrogels under microwave radiations for enhanced thermal and chemical modifications, *Polym. Renew. Resour.*, **2**, pp. 105–116.

27. Sworn, G. (2000). Xanthan Gum, in *Handbook of Hydrocolloids* (Phillips, G. O., and Williams, P. O., eds), CRC Press, New York.

28. Kuo, M. S., Mort, A. J., and Dell, A. (1986). Identification and location of L-glycerate, an unusual acyl substituent in gellan gum, *Carbohydr. Res.*, **156**, pp. 173–187.

29. Brown, J. C., and Livesey, G. (1994). Energy balance and expenditure while consuming Guar gum at various fat intakes and ambient temperatures, *Am. J. Clinic. Nutr.*, **60**, pp. 956–964.

30. Soppirnath, K. S., Kulkarni, A. R., and Aminabhavi, T. M. (2000). Controlled release of antihypertensive drug from the interpenetrating network poly(vinyl alcohol)-guar gum hydrogel microspheres, *J. Biomed. Sci.*, **11**, pp. 27–43.

31. Chauhan, K., Chauhan, G. S., and Ahn, J. H. (2009). Synthesis and characterization of novel guar gum hydrogels and their use as Cu^{2+} sorbents, *Bioresource Technol.*, **100**, pp. 3599–3603.

32. Singh, V., Kumari, P., Pandey, S., and Narayan, T. (2009). Removal of chromium (VI) using poly(methylacrylate) functionalized guar gum, *Bioresource Technol.*, **100**, pp. 1977–1982.

33. Morrison, N. A., Sworn, G., Clark, R. C., Talashek, T., and Chen, Y. L. (2002). New Textures with High Acyl Gellan Gum, in *Gum and Stabilisers for the*

Food Industry 11 (Williams, P. O., and Phillips, G. O., eds), RSC Special Publication, Cambridge, pp. 297–305.

34. Cronin, C. E., Giannouli, P., McCleary, B. V., Brooks, M., and Morris, E. R. (2002). Formation of strong gels by enzymatic debranching of guar gum in the presence of ordered xanthan, in *Gum and Stabilisers for the Food Industry 11* (Williams, P. O., and Phillips, G. O., eds), RSC Special Publication, Cambridge, pp. 286–296.

35. Torres, M. D., Moreira, R., Chenlo, F., and Vázquez, M. J. (2012). Water adsorption isotherms of carboxymethyl cellulose, guar, locust bean, tragacanth and xanthan gums, *Carbohydr. Polym.*, **89**, pp. 592–598.

36. Martins, T., Cerqueira, M. A., Bourbon, A. C., Pinheiro, A. C., Souza, B. W. S., and António, A. V. (2012). Synergistic effects between k-carrageenan and locust bean gum on physicochemical properties of edible films made thereof Joana, *Food Hydro.*, **29**, pp. 280–289.

37. Aspinall, G. O., Hirst, E. L., and Wickstrom, A. (1955). Gum ghatti (Indian gum). The composition of the gum and the structure of two aldobiouronic acids derived from it, *J. Chem. Soc.*, pp. 1160–1165.

38. Preiss, J. (1980). *Chemistry of Cell Wall Polysaccharides*, Academic Press, New York.

39. Meer, G. M., and Gerard, T. (1973). *Industrial Gums*, Academic Press, New York.

40. Kaith, B. S., Jindal, R., Mittal, H., and Kumar, K. (2012). Synthesis, characterization and swelling behavior evaluation of Gum ghattiand acrylamide-based hydrogel for selective absorption of saline from different petroleum fraction–saline emulsions, *J. Appl. Polym. Sci.*, **124**, pp. 2037–2047.

41. Hoffman, A. S., Afrassiabi, A., and Dong, L. C. (1986). Thermally reversible hydrogels: II. Delivery and selective removal of substances from aqueous solutions, *J. Control. Release*, **4**, pp. 213–222.

42. Kuhn, W., Hargitay, B., Katchalsky, A., and Eisenberg, H. (1950). Reversible dilation and contraction by the state of ionization of high-polymer acid networks, *Nature*, **165**, pp. 514–515.

43. Kikuchi, A., and Okano, T. (2002). Intelligent thermoresponsive polymeric stationary phases for aqueous chromatography of biological compounds, *Prog. Polym. Sci.*, **27**, pp. 1165–1193.

44. Yokoyama, M. (2002). Gene delivery using temperature-responsive polymeric carriers, *Drug Discov. Today*, **7**, pp. 426–432.

45. Nandkumar, M. A., Yamato, M., Kushida, A., Konno, C., Hirose, M., Kikuchi, A., and Okano, T. (2002). Two-dimensional cell sheet manipulation of

heterotypically co-cultured lung cells utilizing temperature-responsive culture dishes results in long-term maintenance of di differentiated epithelial cell functions, *Biomaterials*, **23**, pp. 1121–1130.

46. Guenet, J. M. (1992). *Thermoreversible Gelation of Polymers and Biopolymers*, Academic Press, London.

47. Xu, F. J., Kang, E. T., and Neoh K. G. (2006). pH- and temperature-responsive hydrogels from cross-linked triblock copolymers prepared via consecutive atom transfer radical polymerization, *Biomaterials*, **27**, pp. 2787–2797.

48. Chen, J. P., and Cheng, T. H. (2009). Preparation and evaluation of thermo-reversible copolymer hydrogels containing chitosan and hyaluronic acid as injectable cell carriers, *Polymer*, **50**, pp. 107–116.

49. Fujishige, S., and Ando, K. K. I. (1989). Phase transition of aqueous solutions of poly(N-isopropylacrylamide) and poly(N-isopropylmethacrylamide), *J. Phys. Chem.*, **93**, pp. 3311–3313.

50. Zhang, H. F., Zhong, H., Zhang, L. L., Chen, S. B., Zhao, Y. J., and Zhu, Y. L. (2009). Synthesis and characterization of thermosensitive graft copolymer of N-isopropylacrylamide with biodegradable carboxymethylchitosan, *Carbohydr. Polym.*, **77**, pp. 785–790.

51. Chuang, C. Y., Don, T. M., and Chiu, W. Y. (2009). Synthesis and properties of chitosan-based thermo- and pH-responsive nanoparticles and applications in drug release, *J. Polym. Sci., Part A. Polym. Chem.*, **47**, pp. 2798–2810.

52. Gil, E. S., and Hudson, M. (2004). Stimuli-reponsive polymers and their bioconjugates, *Prog. Polym. Sci.*, **29**, pp. 1173–1222.

53. Bajpai, S. K. (1999). Casein cross-linked polyacrylamide hydrogels: Study of swelling and drug release behaviour, *Iran. Polym. J.*, **8**, pp. 231–239.

54. Mittal, H., Kaith, B. S., and Jindal, R. (2010). Microwave radiation induced synthesis of Gum ghatti and acrylamide based cross-linked network and evaluation of its thermal and electrical behaviour, *Der. Chem. Sin.*, **3**, pp. 59–69.

55. Mahdavinia, G. R., Pourjavadi, A., Hosseizadeh, H., and Zohuriaan, M. J. (2004). Modified chitosan 4. Superabsorbent hydrogels from poly(acrylic acid-coacrylamide) grafted chitosan with salt- and pH-responsiveness properties, *Euro. Polym. J.*, **40**, 1399–1407.

56. Berlin, A. A., and Kislenko, V. N. (1992). Kinetics and mechanism of radical graft polymerization of monomers onto polysaccharides, *Prog. Polym. Sci.*, **17**, pp. 765–825.

57. Sui, X., Yuan, J., Zhou, M., Zhang, J., Yang, H., and Yuan, W. (2008). Synthesis of cellulose-graft-poly(N,N-dimethylamino-2-ethyl methacrylate) copolymers via homogeneous atrp and their aggregates in aqueous media, *Biomacromolecules*, **9**, pp. 2615–2620.

58. Ghorai, S., Sinhamahpatra, A., Sarkar, A., Panda, A. B., and Pal, S. (2012). Novel biodegradable nanocomposite based on XG-g-PAM/SiO$_2$: Application of an efficient adsorbent for Pb^{2+} ions from aqueous solution, *Bioresource Technol.*, **119**, pp. 181–190.

59. Arshady, R. (1990). Development of new hydrophilic polymer supports based on dimethylacrylamide, *Colloid. Polym. Sci.*, **268**, pp. 948–958.

60. Arshady, R. (1992). Suspension, emulsion, and dispersion polymerization: A methodological survey, *Colloid. Polym. Sci.*, **270**, pp. 717–732.

61. Sugihara, S., Ohashi, M., and Ikeda, I. (2007). Synthesis of fine hydrogel microspheres and capsules from thermoresponsive coacervate, *Macromolecules*, **40**, pp. 3394–3401.

62. Du, J., Sun, R., Zhang, L. F., Zhang, L. F., Xiong, C. D., and Peng, Y. X. (2005). Novel polyelectrolyte carboxymethyl konjac glucomannan–chitosan nanoparticles for drug delivery. I. Physicochemical characterization of the carboxymethyl konjac glucomannan–chitosan nanoparticles, *Biopolymers*, **78**, pp. 1–8.

63. Sarmento, B., Ribeiro, A., Veiga, F., and Ferreira, D. (2006). Development and characterization of new insulin containing polysaccharide nanoparticles, *Colloids Surf. B*, **53**, pp. 193–202.

64. Rio, S., and Delebarre, A. (2003). Removal of mercury in aqueous solution by fluidized bed plant fly ash, *Fuel*, **82**, pp. 153–159.

65. Cassano, A., Molinari, R., Romano, M., and Drioli, E. (2001). Treatment of aqueous effluents of the leather industry by membrane processes: A review, *J. Membrane Sci.*, **181**, pp. 111–126.

66. Tripathy, J., Mishra, D. K., Srivastava, A., Mishra, M. M., and Behari, K. (2008). Synthesis of partially carboxymethylated guar gum-g-4-vinyl pyridine and study of its water swelling, metal ion sorption and flocculation behavior, *Carbohydr. Polym.*, **72**, pp. 462–472.

67. Turk, S. S., and Reinhold, S. (2000). Printing properties of a high substituted guar gum and its mixture with alginate, *Dyes Pigments*, **47**, pp. 269–275.

68. Lehto, J., Vaaramaa, K., Vesterinen, E., and Tenhu, H. (1998). Uptake of zinc, nickel, and chromium by N-isopropyl acrylamide polymer jells, *J. Appl. Polym. Sci.*, **68**, pp. 355–362.

69. Sand, A., Yadav, M., Mishra, M. M., Tripathy, J., and Behari, K. (2011). Studies on graft copolymerization of 2-acrylamidoglycolic acid onto partially carboxymethylated guar gum and physicochemical properties, *Carbohydr. Polym.,* **83**, pp. 14–21.
70. Janakia, V., Vijayaraghavanb, K., Ohc, B. T., Leec, K. J., Muthucheliand, K., Ramasamya, A. K., and Kannanc, S. K. (2012). Starch/polyaniline nanocomposite for enhanced removal of reactive dyes fromsynthetic effluent, *Carbohydr. Polym.,* **90**, pp. 1437–1444.

Chapter 3

A View on Cellulosic Nanocomposites for Treatment of Wastewater

D. Saravana Bavan and G. C. Mohan Kumar
Department of Mechanical Engineering,
National Institute of Technology Karnataka,
Srinivasnagar, Surathkal, Mangalore, India
saranbav@gmail.com

Natural fibers are low cost, abundant, and biodegradable. Application of natural fibers and their composites in wastewater treatment is a fascinating idea. Generally wastewater treatments include physical, chemical, biological, and mechanical methods with primary, secondary, and tertiary procedures. This chapter is mainly focused on recent survey on technology, preparation, and improved methods in utilizing cellulosic nanocomposites for handling organic waste materials. The work is highlighted on using natural fibers in obtaining activated carbon fibers for adsorbing waste particles and can be used for the treatment of wastewater. High porosity and structures are helpful in trapping low molecular weight chemicals and organic compounds. Cellulosic nanoparticles such as cellulose nanowhiskers, nanocrystalline cellulose, and nanofibers are used as reinforcement in polymer matrices. These renewable nanoparticles are used for the treatment of wastewater.

3.1 Introduction

Cellulose-based materials are playing a major role in composites and films because they have almost superior properties compared to traditional materials. These materials have added advantages such as renewability, biodegradability, low density, low cost, ease of availability, less energy consumption, high specific strength, high modulus, and nonabrasive [67, 106, 110, 117]. Sources of cellulose materials include plants, trees, bacteria, and animals. Plant sources include hemp, cotton, and wood fibers, and others are found in bacteria and tunicates. Naturally occurring cellulosic fibers, fibrils, particles, and crystals are found to be a good reinforcement and filler material for composites [42, 90, 93]. They have added mechanical properties that can compete with man-made fibers of glass and carbon and aramid fibers. The properties of natural fibers are mostly influenced by physical and chemical composition, cellulose content, degree of polymerization, and structure and cross section of fibers.

The present work is focused on surveying various cellulosic nanofiber composites in handling wastewater and improving the condition of water with the treatment of cellulose nanofibers. The role of cellulosic nanocrystals and fibers is appreciated in handling wastewater.

3.2 Classification of Natural Fibers

Natural fibers are composites because they are composed of a combination of cellulose, hemicellulose, lignin, pectin, and wax. They are derived from leaf (e.g., sisal), bast (e.g., flax, hemp), seed (e.g., cotton), and fruit (e.g., coir) but also from other sources such as chicken feathers. Generally, natural fibers are classified into three categories: vegetable, animal, and mineral fibers; among them, mineral fibers are no longer or very rarely used due to their carcinogenic effect. All cellulosic fibers such as cotton, flax, jute, and hemp contain mainly cellulose and hemicellulose. Animal fibers are obtained from animal origin such as hair, silk, and wool, which mainly contains proteins [67, 106, 110, 117]. Vegetable (cellulose) fibers are further classified into bast, leaf, or seed fibers, according to their origin, as shown in Fig. 3.1. From an environmental perspective, natural fibers are biodegradable and are carbon positive since they absorb more carbon dioxide than they produce [109].

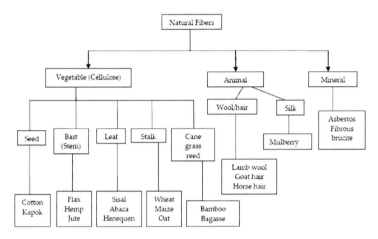

Figure 3.1 Classification of natural fibers. From refs. [67, 110].

Composite materials have a bulk phase, which is continuous, called the matrix, and one dispersed, noncontinuous phase, called the reinforcement, which is usually harder and stronger. The reinforcement material can be of fibers, particulates, or flakes. The concept of composites is that the bulk phase accepts the load over a large surface area and transfers it to the reinforcement, which, being stiffer, increases the strength of the composite [60, 103].

In cellulosic fiber reinforced composites, fiber acts as reinforcement and exhibits high tensile strength and stiffness, and the main purpose of the matrix is to distribute the load to the fibers and act as a protective agent [182]. It also serves to improve impact and fracture resistance of the component and holds the fibers together. The bast and leaf fibers lend mechanical support to the plant's stem and leaf, respectively [80, 124, 145]. The surfaces of natural fibers are uneven and rough, which provides good adhesion to the matrix in a composite material.

Natural fibers are becoming popular in recent times due to vast applications in composite industries. They have a lot of advantages in terms of low cost, availability, processing features, low density, biodegradability, and other favorable mechanical properties [27, 147]. It has become a good competitor for traditional fibers in the composites sector [69]. Natural fibers are widely used in the applications of building structures (paneling, partition board, ceiling), automobiles, packaging industries, and electronics sector [56, 58, 102].

3.3 Structure of Natural Fibers

Natural fibers are generally lignocellulosic in nature, consisting of helically wound cellulose microfibrils in a matrix of lignin and hemicellulose. Cellulose is a polymer containing glucose units and is the main structural component that provides strength and stability to the plant cell walls and the fiber [18, 168]. Hemicellulose is a polymer made of various polysaccharides and the molecules are hydrogen bonded to cellulose and act as cementing matrix between the cellulose microfibrils, forming the cellulose–hemicellulose network [15, 144]. Lignin is an amorphous and heterogeneous mixture of aromatic polymers and phenyl propane monomers. The hydrophobic lignin network affects the properties of other networks in a way that it acts as a coupling agent and increases the stiffness of the cellulose/hemicellulose composite [15].

The cell wall in a fiber is not a homogenous membrane [168]. Each fiber has a complex, layered structure consisting of a thin primary wall, which is the first layer deposited during cell growth encircling a secondary wall. The secondary wall is made up of three layers, and the thick middle layer determines the mechanical properties of the fiber. The middle layer consists of a series of helically wound cellular microfibrils formed from long-chain cellulose molecules. The angle between the fiber axis and the microfibrils is called the microfibrillar angle [18]. Some of the plant fiber microfibril angle is shown in Table 3.1.

The characteristic value of microfibrillar angle varies from one fiber to another, and microfibrils have a typical diameter of about 10–30 nm and are made up of 30–100 cellulose molecules in extended chain conformation and provide mechanical strength to the fiber [18, 67, 106, 168, 169]. The amorphous matrix phase in a cell wall is very complex and consists of hemicellulose, lignin, and in some cases pectin.

Table 3.1 Plant fiber with microfibril angle

Plant fiber	Microfibril angle (degrees)
Hemp	6
Flax	6–10
Jute	8
Sisal	10–25

Source: [67, 106].

Table 3.2 shows fiber sources and its origin. Natural fibers as reinforcement in composite materials have gained new interests. Indeed, ecological considerations such as recyclability and environmentally friendly products are the new driving forces in our society where pollution and global warming issues have become almost uncontrollable [72, 95].

Table 3.2 Some fiber sources and their origin

Fiber source	Species	Origin
Abacca	Musa textiles	Leaf
Banana	Musa indica	Leaf
Cotton	Gossypium sp.	Seed
Coir	Cocos nucifera	Fruit
Flax	Linum usitatissum	Stem
Hemp	Cannabis sativa	Stem
Jute	Corchorus capsularis	Stem
Kenaf	Hibiscus cannabinus	Stem
Pineapple	Ananus comosus	Leaf
Ramie	Boehmeria nivea	Stem
Sisal	Agave sisalana	Leaf

Source: [67, 106].

3.4 Physical, Mechanical, and Other Properties of Natural Fibers

Natural fibers possess a number of advantages in terms of specific material properties as shown in Table 3.3. For light-weight purposes, substitution of synthetic fibers with natural fibers can reduce the weight of a composite by up to 40%, thereby offering substantial gains in fuel efficiency for the automotive, transport, and aerospace sectors [109]. Improvements in flexural strength, stiffness, and ductility can be achieved through the substitution of natural fibers [57].

Table 3.3 Properties of natural fibers and synthetic fibers

Fiber	Density (g/cm^3)	Tensile strength (MPa)	Young's modulus (GPa)	Elongation at break
Cotton	1.5–1.6	287–800	5.5–12.6	7.0–8.0
Jute	1.3–1.45	393–773	13–26.5	1.16–1.5
Flax	1.50	345–1100	27.6	2.7–3.2
Sisal	1.45	468–640	9.4–22.0	3–7
Coir	1.15	131–175	4–6	15–40
Ramie	1.50	400–938	61.4–128	1.2–3.8
E-glass	2.5	2000–3500	70	2.5
Carbon	1.7	4000	230–240	1.4–1.8

Source: [67, 106, 110, 117].

The ultimate properties of the composites depend on many properties of the constituents, size and shape of the individual reinforcing fibers or particles, structural arrangement and distribution, relative amount of each constituent, and the interface between matrix and reinforcement [81, 89]. Composite mechanical properties are strongly influenced by the mechanical properties, distribution of the fibers/matrix, and from the efficiency stress transfer between these two components [13, 20, 22]. The mechanical properties of reinforcement (fibers) have direct relation with the tensile strength and stiffness of the composite [43, 44, 52, 76, 85, 135]. The selection of suitable reinforcing fibers follows certain criteria such as thermal stability, fiber–matrix adhesion, long-time behavior, elongation at failure, price, and processing costs [48, 67, 119, 121].

3.4.1 Problems with Natural Fibers

One of the major problems associated with the use of natural fibers in composites is their high moisture sensitivity, leading to severe reduction of mechanical properties [7, 37, 94, 143, 179]. The reduction may be due to poor interfacial bonding between resin matrices and fibers. It is, therefore, necessary to modify the fiber surface to render it more hydrophobic and also more compatible with resin matrices [51, 53, 77, 83]. Generally, acids and alkali have been used for modifying jute, coir, and other natural fibers [160, 181]. Strong alkali solutions lead to reduction of strength, an increase in elongation at break, and reduction in stiffness of the fibers; whereas lower concentration of alkali (10%) does not cause significant lowering in strength [4, 105, 157]. The influences of surface treatments in natural fibers on the interfacial characteristics were studied [66, 108]. The micromechanics approach and the thermo-elastic anisotropy [29] of natural fiber composites were also determined by researchers.

The water sorption capacity of plant fibers is an essential aspect of plant fiber composites [3, 40, 67, 100, 110, 136, 164]. Understanding water sorption in the composites requires necessarily an understanding of water sorption in the fibers themselves [3]. The affinity between liquid water and a solid material is characterized by the contact angle at the water–material interface, and the material is denoted hydrophilic when the angle is below 90° and hydrophobic when the angle is above 90°.

3.4.2 Limitations of Natural Fibers

The limitations of natural fibers are as follows [12, 67, 101, 106, 112, 150, 155]:
- Lower strength properties
- Poor fire resistance
- Lower durability

- Variability in quality, depending on unpredictable influences such as weather
- Moisture absorption, which causes swelling of the fibers
- Restricted maximum processing temperature (up to 220 °C)
- Fluctuation of price depending on harvest results
- Labor-intensive processing of fiber
- Dimensional stability

3.5 Chemical Composition of Natural Fibers

The cell wall of plant fibers is mainly composed of polymers: cellulose, hemicelluloses, lignin, and pectin. These polymers differ in molecular composition and structure and, therefore, display different mechanical properties as well as different water sorption properties [163]. The content of the polymers is highly variable between plant fibers. Table 3.4 shows the chemical composition of natural fibers.

Table 3.4 Chemical composition of natural fibers

Fiber type	Cellulose	Hemicellulose	Lignin	Pectin
Abaca	61–64	21	12	0.8
Banana	60–65	6–19	5–10	3–5
Bamboo	26–43	15–26	21–31	—
Cotton	82–96	2–6	0.5–1	5–7
Flax	60–81	14–19	2–3	0.9
Hemp	70–92	18–22	3–5	0.9
Jute	51–84	12–20	5–13	0.2

Source: [67, 110].

3.5.1 Cellulose

Cellulose is the main component of the plant cell wall, with a share of 60–90%. It is a linear condensation polymer consisting of D-anhydro glucopyranose units joined together by β-1,4-glycosidic bonds. Glucose is bonded to the next glucose through 1 and 4 carbons as shown in Fig. 3.2 to form cellulose. The average length of a cellulose chain is 5000 cellulose units (5 μm). These long, flat chains can bond tightly together to form crystalline regions [36, 153]. The molecular structure of cellulose is responsible for its supramolecular structure, and this determines many of its chemical and physical properties [153]. The mechanical properties of natural fiber depend on its cellulose type because each type of cellulose has its own cell geometry, which determines the chemical properties.

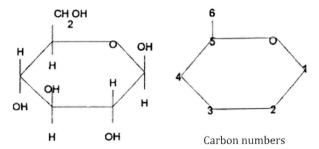

Figure 3.2   Cellulose chemical structure.

Cellulose content is an important parameter, because in chemical pulping, the pulp yield corresponds to the cellulose content of the raw material. At the same time, it is the main strengthening element of this natural composite material [11]. The mechanical properties are quite remarkable: Cellulose filament of 1 mm diameter can hold more than 60 kg in weight, and it has 80% of the strength of steel [70].

3.5.2   Hemicellulose

Hemicelluloses are derived mainly from chains of pentose sugars and act as the cement material holding together the cellulose and the fiber [34, 72]. They are made up of chains of sugars and comprise a group of polysaccharides (excluding pectin) bonded together in relatively short, branching chains and remain associated with the cellulose after lignin has been removed. These sugars include glucose and other monomers such as galactose, mannose, xylose, and arabinose as shown in Fig. 3.3.

Hemicellulose differs from cellulose in three important aspects. First, it contains several different sugar units, whereas cellulose contains only 1,4-β-D-glucopyranose units. Second, it exhibits a considerable degree of chain branching, whereas cellulose is strictly a linear polymer. Last, the degree of polymerization of native cellulose is between 10 and 100 times higher than that of hemicellulose, and hemicellulose cannot pack together as tightly as cellulose, and the constituents of hemicellulose differ from plant to plant [59, 72]. The backbone chains of hemicelluloses can be a homopolymer (generally consisting of single sugar repeat unit) or a heteropolymer (mixture of different sugars).

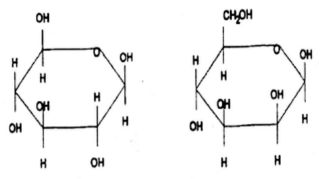

Figure 3.3 Hemicellulose chemical structure.

3.5.3 Lignin

Lignin is the compound that gives rigidity to fiber. It is a complex hydrocarbon polymer with both aliphatic and aromatic constituents. Its main monomer units are the various ring-substituted phenylpropane linked together in ways [67, 106]. Lignin is one of the major constituents of wood and is the extractive component of wood that binds the cellulose fibers together and removed during the pulping process.

The main chemical difference between lignin, cellulose, and hemicellulose is the amount of potential cross-linking sites on lignin. Cross-linking can occur along the propane chain, through the C4 oxygen and at the vacant aromatic ring carbons. The cross-linking of lignin gives an irregular amorphous (noncrystalline) structure. An extracted lignin is often used as a phenol substitute in matrix materials. Functions of lignin include providing structural strength, providing sealing of water conducting system that links roots with leaves, and protecting plants against degradation [106].

3.5.4 Pectin and Others

Pectin is a collective name for heteropolysaccharides, which consists essentially of polygalacturon acid. They are complex polysaccharides, and the main chains consist of a modified polymer of glucuronic acid and residues of rhamnose. The chains are rich in rhamnose, galactose, and arabinose sugars [67, 106]. These chains are often cross-linked by calcium ions, improving structural integrity in pectin-rich areas. Pectin is important in nonwood fibers, especially bast fibers. Lignin, hemicelluloses, and pectin collectively function as matrix and adhesive, helping to hold together the cellulosic framework structure of the natural composite fiber. Pectin is soluble in water only after a partial neutralization with alkali.

The waxes make the part of the fibers, which can be extracted with organic solutions. Waxy materials consist of different types of alcohols, which are insoluble in water and several acids such as palmitic acid and oleaginous acid.

3.6 Biocomposites/Green Composites

Bio-based composite materials are a new and innovative class of materials being developed today and consist of environmentally friendly resins and natural fibers [35, 124, 134]. Generally, natural fiber acts as reinforcement material, and the matrix material can be of synthetic polymer or biopolymer [6, 107, 127, 175, 176, 180]. Natural fibers from vegetable are lignocellulosic fibers, where the cellulose provides the strength while the lignin and hemicellulose provide the toughness and protection of the fibers [68]. Single fibers are themselves made of several microfibrils. A good orientation angle of these microfibrils, as well as high cellulose content, gives better mechanical properties. Mechanical separation of fibers is carried out using decorticating machines, steam explosion, and ammonia fiber extraction [32, 46, 138, 139]. Retting is the traditional process to extract fibers using bacteria and fungi in the environment to remove lignin, pectin, and other substances. More common chemical retting methods use alkalis, mild acids, and enzymes for fiber extraction [159]. Life cycle assessment of biofibers replacing glass fibers was also studied and proved that biofibers are equivalent to glass fibers at some specific conditions and methods [69].

3.7 Wastewater Treatment

Wastewater is composed of organic and inorganic compounds. Organic compound components consist of proteins, carbohydrates, fats, greases, phenols, oils, surfactants, and pesticides. Inorganic compound components consist of nitrogen, phosphorus, chlorides, and heavy metals. Wastewater also contains various microorganisms such as bacteria, fungi, algae, and protozoa. Wastewater resources are from two sources, namely, domestic wastewater and industrial wastewater. Domestic wastewater (sewage water) includes waste from bathroom, kitchen, laundry, and also includes discharged water from institutional and commercial sources. Heavy metals such as cadmium, chromium, copper, mercury, lead, and zinc are present in wastewater from industries [19, 24]. General wastewater treatment is shown in Fig. 3.4. Requirements of costly equipment, high capital, operational costs, monitoring system, high reagent, and energy requirements are expected drawbacks from the conventional water treatment systems.

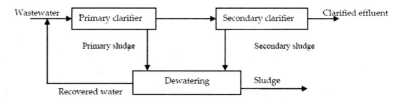

Figure 3.4 Conventional wastewater treatment. From ref. [156].

Textile industries are the major one in polluting water sources because of chemical processing, dyeing, printing, finishing, and other processing features [84, 128, 166]. The water contains organic and inorganic materials, which include surfactants, toxic compounds, phosphates, and pesticides as shown in Table 3.5.

Table 3.5 Distribution of heavy metals in few industrial effluents

Industries	As	Cd	Cr	Cu	Fe	Hg	Mn	Ni	Pb	Ti	Zn
Fertilizers		√	√	√	√	√	√	√	√		√
Pesticides	√			√		√					
Tanning	√		√								

Industries	As	Cd	Cr	Cu	Fe	Hg	Mn	Ni	Pb	Ti	Zn
Dyeing			√						√		
Paper products			√	√		√		√	√	√	√
Plating		√	√	√				√	√		√
Cooling water				√							

Source: [25].

3.8 Classification of Wastewater Treatment

Generally wastewater treatment involves three stages: primary, secondary, and tertiary [25, 87, 115]. Primary treatments are used to reduce grease, fats, oils, sand, and grit solids and are carried out by means of filtration and sedimentation. The treatment includes removing the suspended small or large particles and skimming oil from the wastewater. Secondary treatment is used to degrade the organic content of the sewage, and often microorganisms are also used in the purification, which convert nonsettable solids to settable solids. Biological treatment is used for the reduction and removal of organic contaminants as shown Fig. 3.5. Classification of wastewater treatment is shown in Table 3.6 [25, 115, 122].

- Primary treatment includes screening, grit chambers, sediment tank, aeration, and filtration
- Secondary treatment includes activated sludges, tertiary filters, and lagoons
- Tertiary treatment includes physicochemical treatment (using activated carbon), flocculation, coagulation, membrane filtration, hyper filtration

Wastewater treatment can also be classified into physical, chemical, and biological methods [25, 115] as shown in Table 3.7.

- Physical methods include sedimentation, screening, aeration, filtration
- Chemical methods include chlorination, ozonation, neutralization, coagulation, adsorption, ion exchange
- Biological methods include aerobic digestion, activated sludge treatment, trickling filtration, oxidation ponds,

and lagoons (aerobic), and anaerobic digestion and septic tanks (anaerobic)

Table 3.6 Classification of wastewater treatment

Treatment	Operations
Primary	Screening
	Sedimentation
	Equalization
	Mechanical flocculation
	Chemical coagulation
Secondary	Aerated lagoon
	Trickling filtration
	Activated sludge process
	Anaerobic digestion
Tertiary	Oxidation technique
	Electrolytic precipitation
	Ion exchange method
	Photocatalytic degradation
	Adsorption
	Thermal evaporation

Source: [115].

Table 3.7 Physical, chemical, and biological water treatments

		Biological	
Physical	Chemical	Aerobic	Anaerobic
Screening	Chlorination	Activated sludge treatment methods	Anaerobic digestion
Aeration	Ozonation	Trickling filtration	Septic tanks
Filtration	Neutralization	Oxidation ponds	Lagoons
Flotation and skimming	Coagulation	Lagoons	
Degasification	Adsorption	Aerobic digestion	
Equalization	Ion exchange		

Source: [25, 115].

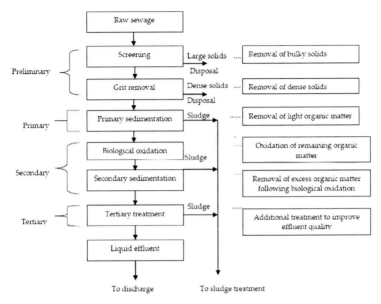

Figure 3.5 Typical stages in conventional treatment of sewage. From ref. [146].

Common operations used in wastewater treatment include screening, grit removal, sedimentation, coarse solids reduction, flow equalization, mixing and flocculation, high-rate clarification, accelerated gravity separation, flotation, volatilization and stripping of volatile organic compounds, oxygen transfer, packed-bed filtration, membrane separation, and aeration.

Various separation processes with different treatments and their application are shown in Table 3.8.

3.9 Dye in Wastewater

Dyes are usually found in many industries such as textiles, paper, printing, leather, and others. They are used to impart color to other materials and are absorbed when applied. Sometimes fast dye and fugitive dye terms are used, when dye used are stable or unstable (losing their colors). Sunlight, exposure to acids, excessive heat, and washing procedures are some of the considerations that can cause change in the dye [87, 88].

Table 3.8 Representative unit processes and operations for water reclamation

	Process	Description	Application
Solid/liquid separation	Sedimentation	Gravity sedimentation of particulate matter and precipitates from suspension-gravity settling	Removal of particles greater than 30 µm from turbid water
	Filtration	Particle removal by passing water through sand or other porous substance	Removal of particles greater than 3 µm from water. Used after sedimentation or coagulation
Biological treatment	Aerobic biological treatment	Biological metabolism of wastewater by microorganisms	Removing of dissolved and suspended organic matter
	Biological nutrient removal	Combination of aerobic and anaerobic processes	Reducing nutrient content of water
Advancement treatment	Activated carbon	Contaminants are physically adsorbed on the surface carbon	Removal of hydrophobic organic compounds
	Ion exchange	Exchange of ions between exchange resin and water	Removal of cations and anions
	Membrane Filtration	Micro-, nano-, and ultrafiltration	Removal of particles and microorganisms
	Reverse osmosis	Membrane system to separate ions from solutions	Removal of dissolved salts and minerals

Source: [173].

Treatment methods of dye containing wastewater include chemical oxidation, photo degradation, reverse osmosis, adsorption, electrocoagulation and electroflotation. Advantages and disadvantages of dye removal are listed in Table 3.9 and Table 3.10.

Table 3.9 Existing processes of dyes removal and their advantages

Treatment process	Technology	Advantages
Conventional treatment process	Coagulation, Flocculation	Simple, economically feasible
	Biodegradation	Economically attractive, acceptable treatment
	Adsorption on activated carbons	Effective adsorbent, great, capacity, produce high quality treated effluent
Established recovery process	Membrane separations	Removes all dye types, produces high quality treated effluent
	Ion exchange	No loss of sorbent on regeneration, effective
	Oxidation	Rapid and efficient process
Emerging removal process	Advanced oxidation process	No sludge production, no consumption of chemicals

Source: [30, 31].

Table 3.10 Disadvantages of conventional methods for dye removal

Treatment process	Technology	Disadvantages
Conventional treatment process	Coagulation, flocculation	High sludge production, problems in handling and disposal
	Biodegradation	Slow process, nutrition requirements
	Adsorption on activated carbons	Ineffective against disperse and vat dyes, regeneration is expensive
Established recovery process	Membrane separations	High pressure, expensive, difficult in treating large volumes
	Ion exchange	Economic constraints, not effective for disperse dyes
	Oxidation	High energy cost, chemicals required
Emerging removal process	Advanced oxidation process	Economically unfeasible, technical constraints

Source: [30, 31].

Dyes can be classified according to the use of dying processes, such as acid dye, basic dye, mordant dye, direct dye, reactive dye, vat dye, azoic dye, and disperse dye. The type of dyes and application are shown in Table 3.11.

Table 3.11   Classification of dyes based on their use

Type of dyes	Application
Azo dyes	Cotton
Direct dyes	Cotton, paper, and synthetics
Reactive dyes	Cotton and wool
Sulfur dyes	Cotton and synthetics
Vat dyes	Cotton and synthetics

Source: [25, 96].

3.10 Adsorbents in Wastewater

Toxic ions can be removed from water through various methods and technologies such as adsorption, ion exchange, ion flotation, chemical precipitation, reverse osmosis, and evaporation. In adsorption, a component or impurity in a fluid is removed by contacting the fluid with a solid adsorbent. The adsorbing phase is the adsorbent, and the material adsorbed at the surface of the phase is the adsorbate. Activated carbon is the most widely used adsorbent in industries (Table 3.12). Activated carbon fibers, carbon nanotubes, carbon microspheres, and carbon nanoparticles are used as adsorbents for the aqueous solution for the removal of phenol and metallic ions [25].

Natural adsorbents such as chitosan, clay, zeolite, fly ash, sawdust, fiber sludge ash, fertilizer waste, olive stone, peat moss, pine bark, plant straw, and dried aquatic plants are some of the effectively low cost materials for the treatment of polluted water from heavy metals ions [26]. Waste materials from industrial operations, such as fly ash, coal, and oxides, are also used as low cost adsorbents. Different adsorbents such as activated carbon, anion exchanger, anthracite, polyaniline, and their nanocomposites were used to separate sulfate from industrial wastewater [10].

Table 3.12 Various adsorbents used in wastewater treatment

Adsorbents		Pollutant
Activated carbon		Cr(VI)
		Benzoic acid
		Dyes
		Pb^{2+}, Cd^{2+}
		Cresol
		Chlorophenols
Zeolites	Scolecite	Pb^{2+}, Cu^{2+}, Cd^{2+}
	Clinoptilolite	Pb^{2+}, Cr^{3+}, Cu^{2+}, Fe^{2+}
Low cost adsorbents	Sawdust	Dyes
	Bagasse fly ash	Cd^{2+}
	Sugar beet pulp	Cd^{2+}, Pb^{2+}
	Corncob	Dyes
	Wheat straw	Dyes
	Baker's yeast	Cd^{2+}
	Paper mill sludge	Phenols
Polymeric materials	Organic polymer resin	U (VI), Cu^{2+}
	Hyper cross-linked polymer	Phenols, aniline, benzene
Clays	Montmorillonite	Mn^{2+}, Zn^{2+}, Ni^{2+}
	Bentonite	Phenol
	Sepiolite	Phenol
	Kaolinite	Cu^{2+}, Co^{2+}

Source: [30, 31].

3.10.1 Activated Carbon

Activated carbon is a solid, microcrystalline, nongraphitic form of black carbonaceous material with a porous structure. It is widely used because of the extensive surface area, high adsorption capacity, and high degree of surface reactivity. Using activated carbon to remove metal ions from wastewater by adsorption is one of the best and known processes. Activated carbon materials are characterized by high specific surface area and porosity. The physical properties of activated carbon are shown in Table 3.13.

Table 3.13 Physical properties of adsorbents

Parameter	Neem leaves	Orange peel	Banana peel	Activated carbon
Moisture content (%)	11.10	38.50	7.70	30.00
Ash content (%)	4.30	17.80	5.60	30.00
Volatile content (%)	84.60	43.70	86.70	40.00
Specific gravity	2.85	3.22	3.13	1.83
Fineness modulus	3.128	3.24	3.02	2.17
Particle size (mm)	0.186	0.150	0.200	0.002
Void ratio	0.75	0.75	0.36	0.50
Particle density (g/cc)	0.56	1.44	0.92	1.90

Source: [115].

Activated carbon is also popularly known as activated charcoal. It is a form of carbon processed to make it extremely porous. This carbon is produced from carbonaceous materials such as lignite, coal, petroleum pitch, nut shells, peat, and wood. It can be produced by the following methods:
- Physical reactivation, which includes carbonization and activation/oxidation
- Chemical activation

Activated carbon plays an important role in the adsorption process. It has the ability to remove inorganic and organic waste from industrial and chemical wastes. It is a versatile adsorbent that has a high surface area for adsorption and chemical reactions. The porous structure and high degree of surface reactivity influence its interaction with polar or nonpolar adsorbents. Carbon materials are versatile and unique in performance and have major industrial significance. They are used in removing inorganic and organic pollutants from industrial wastewater, drinking water, decolorizing of syrups, chemicals, and pharmaceutical industries [99]. The properties of activated carbon depend on various factors as listed below [148]:

- Carbon surface area
- Pore size distribution
- Polarity of the surface
- Chemical and porous structure of carbon
- Physical and chemical characteristics of adsorbents

Activated carbon is classified based on their physical characteristics as follows [99, 104]:

- Powdered activated carbon
- Extruded activated carbon
- Granular activated carbon
- Polymer-coated carbon
- Impregnated carbon

Various cellulosic precursors are used for the production of activated carbon, such as Sugarcane bagasse, palm seed, coconut shell, pecan shell, wood, olive stones, and molasses. Among them, coal, wood, and peat are commonly used, and low cost sorbents such as fungi, bacteria, algae, and other lignocellulosic agricultural by-products are used for their biosorption capacity toward heavy metals [64]. Solid wastes, namely, egg plant hull, almond green hull, and moss, are used as adsorbents for the adsorption of strontium ion from aqueous solutions [137].

Activated carbons, plant wastes, clays, and biopolymers are among the common adsorbents used in today's world. Activated carbon–iron oxide magnetic composites are prepared with different weight ratios to produce magnetic adsorbents and are used to remove contaminants in water by magnetic procedure [98]. Chitosan (biopolymer) is used to remove anionic and cationic dyes and metal ions [123]. The capacity of various adsorbents is shown in Table 3.14. Major adsorbents include silica gel, activated alumina, activated carbon, molecular sieve zeolites, and molecular sieve carbon.

Table 3.14 Comparison of adsorbent capacity of various adsorbents

Adsorbent	Q_o (mg/g)
Hydrous Fe oxide with polyacrylamide	43.0
Chemivon F-400 GAC	20.22
Activated alumina	17.61
Alumina	13.64
Activated bauxite	3.89
Activated carbon	1.05
$Al_2O_3/Fe(OH)_3$	0.09

Source: [64].

Activated carbon is used in gold purification, gas purification, water purification, metal extraction, sewage treatment, medicine, and for air filters in gas masks [2, 64]. Activated carbon usage in industries includes the following [2]:

- Water purification in removing phenol, halogenated compounds, pesticides, caprolactum, and chlorine
- Purification of many chemical, food, and pharmaceutical products
- Air purification in inhabited spaces and food processing industries
- Removal of nitrogen from air
- Removal of hydrogen from syngas and hydrogenation process
- Recovery of solvent vapors
- Removal of SO_X and NO_X
- Purification of helium

3.11 Role of Agro-Fibers and Polymers in Handling Wastewater

Agricultural wastes are good alternatives to expensive adsorbents because of their large amount of surface functional groups [125]. Agricultural solid waste (biomass) is a preferred option for low activated carbon precursors, and the availability of other options is too costly. Advantages of biomass sources are low cost, more availability, renewability, and biodegradability. Agricultural wastes from coconut husk, rice husk, wheat straw, barley straw, crofton weed, soybean hull, orange peel, bamboo, sawdust, sunflower, maize cobs, tamarind, banana pith, coir pith, oil palm shell, cashew nut shell, and palm kernel fiber are some of the sources for activated carbon [9, 49, 62, 129, 140, 149, 151, 167, 178]. A few agriculture residues with proximate and ultimate analysis are shown in Table 3.15.

The production of activated carbon involves two steps, namely, carbonization (converting raw materials into solid char and leaving others as by-products) and activation (raising char porosity and cleaning out tar-clogging process). A few crops and by-products for use as adsorbent materials are shown in Table 3.16. Natural fibers are subjected to chemical hydrolysis or to enzymatic hydrolysis in combination with high mechanical shearing forces.

Activation can be done by physical, chemical, and physiochemical methods. Agricultural wastes used for making activated carbon for dye removal include palm tree cobs, plum kernels, and bamboo [74, 171] as shown in Table 3.17. Sugarcane bagasse, olive palm shell, rice husk, and coconut shell are less expensive agricultural adsorbents [118]. These adsorbents contain protein, polysaccharides, and lignin associating with other functional groups and they are responsible for metal ion adsorption [158].

3.12 Biosorption

Biosorption is the capability of active sites on the surface of biomaterials to bind and concentrate heavy metals from even the most dilute aqueous solutions. Biosorption can be used for the treatment of wastewater with low heavy metal concentration [64]. The ability of biological materials to accumulate heavy metals from wastewater is also termed biosorption and is performed through metabolically mediated or physicochemical pathway. They are low cost, inexpensive, simple, highly efficient, and an effective alternative to conventional methods [120, 125]. Agricultural wastes such as rice husk are excellent biosorbents. Rice husks are composed of cellulose, hemicelluloses, lignin, and silica, and these husks contain floristic fiber, protein, and other functional groups (carboxyl, hydroxyl) that make the adsorption processes easier.

Table 3.15 Agricultural residues with proximate and ultimate analysis

Agricultural wastes	Moisture % ww	Ash % ww	Volatiles % ww	C % ww	H % ww	O % ww	N % ww	S % ww	HHV % ww
Cotton stalks	6	13.3	—	41.23	5.03	34	2.63	0	3772
Soft wheat straw	15	13.7	69.8	—	—	—	—	—	4278
Corn stalks	0	6.4	—	45.53	6.15	41.11	0.78	0.13	4253
Corncobs	7.1	5.34	—	46.3	5.6	42.19	0.57	0	4300
Barley straw	15	4.9	—	46.8	5.53	41.9	0.41	0.06	4489
Rice straw	25	13.4	69.3	41.8	4.63	36.6	0.7	0.08	2900
Oats straw	15	4.9	—	46	5.91	43.5	1.13	0.015	4321
Sunflower straw	40	3	—	52.9	6.58	35.9	1.38	0.15	4971
Cherry tree prunings	40	1	84.2	—	—	—	—	—	5198
Peach tree prunings	40	1	79.1	53	5.9	39.1	0.32	0.05	4500

Source: [65].

Table 3.16 Crops and by-products for use as adsorbent materials

Crops	By-products
Karanja	Seed cake
Maize	Stalk, cob
Coconut	Coir, shell
Groundnut	Husk
Melon	Seed husk
Rubber	Seed shell
Rice	Husk

Source: [148].

Table 3.17 Low cost materials for removal of dyes from aqueous solution

Adsorbents	Dyes
Bamboo dust, coconut shell, ground nut shell, rice husk	Methylene blue
Silk cotton hull, coconut tree, sago waste, maize cob	Rhodamine-B, Congo red, methylene blue, malachite green
Rice husk	Malachite green
Coir pith	Acid violet, acid brilliant blue, Rhodamine-B, methylene blue
Banana and orange peels	Methyl orange, Rhodamine-B, methylene blue, methyl violet, Congo red
Banana pith	Congo red, Rhodamine-B, acid violet, acid brilliant blue
Rice husk	Safranine, methylene blue

Source: [74].

3.13 Activated Carbon from Plant Fibers as Adsorbents

Different researchers had contributed toward this subject and some of the studies are summarized here. Adsorption properties of activated carbon prepared from coconut shell was investigated [33], and adsorptions system equilibriums were revealed. Because of hardness and abrasion resistance of coconut shells, further research on them proved effective. Abedi and Bahreini [1] used the plant *Calotropis gigantea* as a material for preparing a carbonaceous adsorbent by thermochemical process. Use of banana pseudo-stem fibers for removal of oil from synthetic wastewater was investigated [61]. These stem fibers exhibit 100% adsorption at lower concentration of oil and had high adsorption capacity. Mundhe et al. [114] studied the use of a sulfuric acid–treated material, which was prepared from *Polyulthia longifolia* seeds, for the removal of methylene blue from water solution. Adsorption studies on single- and multicomponent systems from bagasse waste material were also carried out, as shown in Table 3.18 [104]. Activated carbon was derived, characterized, and utilized for the removal of cadmium and zinc, and uptake of cadmium was found to be greater than that of zinc. Okra wastes were used as an adsorbent for lead removal from aqueous solutions, and concentrations of the adsorbent and the adsorbate were studied using batch adsorption technique [54]. Barley straws were used as a raw material to produce activated carbon [62]. Jute fiber was converted to an effective adsorbent and used for the removal of malathion pesticide from aqueous solution [82].

Table 3.18 Characteristics of activated carbon derived from bagasse

Characteristics	Values
Bulk density (g/cm^3)	0.387
Moisture (%)	7.8
Ash (%)	52
Percent volatile matter	22
pH	7.6
BET surface area (m^2/g)	960
Porosity (%)	68.09
Phenol number (mg/g)	71
Decolorizing power (mg/g)	84
Hardness	< 1

Source: [104].

Mesoporous activated carbon was prepared from coconut coir dust for the adsorption of dye molecules from aqueous solutions

[99]. A kinematic and calorimetric study of adsorption dyes was done. Adsorption of lead (Pb^{2+}) in wastewater by Luffa cylindrical fiber was carried out [154], and different parameters were studied. Jackfruit peel nanoporous adsorbent for removing Rhodamine dye from aqueous solution was also studied [141]. Palm shell was used in preparing activated carbon [45]. Coconut shell charcoal for Cr(VI) removal was investigated [9] using electroplating wastewater and Cr removal performances were improved. Mosambi peel activated carbon was investigated as an adsorbent, and it was revealed that it can be used fruitfully for removing dye from aqueous solutions [86]. Activated carbon from cocoa shell, known as cocoa (*Theobroma cacao*) shell-based activated carbon, was used to adsorb crystal violet dye from aqueous solution [50].

Palm fruit fiber was used for removing Pb, Cu, Ni, and Cr from aqueous solution [63]. Activated carbon from pinewood was investigated [172]. Oil palm shell was used for the adsorption of sulfur dioxide [97]. Lead (II) ion was removed from wastewater through activated carbon produced from fluted pumpkin (*Telfairia occidentalis*) seed shell [126]. Pine sawdust was investigated as an adsorbent in the treatment of wastewater containing cobalt ions [116]. Eucalyptus barks were used for the removal of chromium [152]. Activated carbon prepared from kenaf fiber by using the physiochemical activation method was studied [28]. Flax shive and cotton gin waste was studied as a precursor for activated carbon for the adsorption of trichloroethylene from liquid and gas phases [78]. Activated carbon from banana empty fruit bunches and *Delonix regia* fruit pods was reported [165]. Table 3.19 gives adsorption capacities for several sorbents.

3.14 Cellulose Nanocomposite Materials

Composite materials are generally divided into two classes: fiber-reinforced composites (fibrous) and particle-reinforced composites (particulate). Nanocomposites fall into the particulate composite classification. In cellulose nanocomposites, reinforcement can be in the form of cellulose nanofibers or cellulose nanocrystals. Cellulose nanofibers are also known as cellulose nanofibrils/nanofibrillated cellulose/microfibrillated cellulose.

Table 3.19 Adsorption capacities (mg/g) for several sorbents

Material	Cd	Cr	Hg	Pb	Ni	Zn	Cu
Dry pine needles			175				
Dry red wood leaves			175				
Undyed jute			7.6	7.9			
Undyed bamboo pulp			9.2	8.4			
Orange peel		12					
Senna leaves		25					
Unmodified jute					3.37	3.55	4.23
Papaya wood	17.35					14.44	19.99
Lignocelluloses fiber					7.49	7.88	
Wheat shell							10.84
Peanut hulls					30	9	8
Peanut hull pellets	6					10	10
Activated carbon from baggase	49.07					14.0	
Carrot residue		45.09				29.61	32.74

Source: [25].

Cellulose is the most abundant natural biopolymer on earth, which is renewable, biodegradable, as well as nontoxic [111]. Cellulose is found in plant fibers such as cotton, wood, hemp, coconut, and other sources from bacteria and tunicates. Cellulose is in the form of microfibrils surrounded by hemicellulose, lignin, pectin, and other wax in the cell wall of plants [16]. Hemicelluloses and lignin in organics and extractives must be extracted and cellulose chains deconstructed to extract cellulose nanowhiskers from plant fibers [41].

Cellulose fibers consist of crystalline and amorphous regions, whereas amorphous regions have lower density compared to crystalline regions. When cellulose fibers are subjected to acid treatment (acid hydrolysis), the amorphous regions break up releasing individual crystallites [133].

Cellulosic materials are used as nanomaterials because they have a nanofibrillar structure and also can be made multifunctional [177]. Nanowhiskers can be used as reinforcement in nanocomposites, which can be obtained by acid hydrolysis from sources such as cotton, sugar beet, wheat straw, ramie, wood, tunicin, and bacterial cellulose [16, 91]. The properties of nanocomposite materials depend not only on the properties of their individual but also on their morphology and interfacial characteristics [39]. Cellulose-based microfibrillated celluloses are of much use in composites, coatings, and films because of their mechanical properties and abundant fiber precursor [79, 174]. Microfibrillated cellulose films from pulps containing varying amounts of hemicelluloses, lignin, and extractives were produced by homogenization techniques [16]. The properties of nanocrystalline cellulose depend on different factors such as type of acid used for hydrolysis, cellulose sources, reaction time, and temperature.

Cellulosic waste materials such as corncob, sugarcane bagasse, sawdust, and wastepaper were converted into nanocomposites and nanostructured ceramics [16, 131], and their effect on cellulosic materials on the properties was found. Nanocomposite films of bacterial cellulose and polyurethane resin were prepared and their properties were characterized [71]. Cellulose nanofibril reinforced polycarbonate composites by compression molding was carried out [132]. These nanofibers were prepared from wood pulp fibers by mechanical defibrillation and characterized for strength properties.

3.15 Cellulose Nanocrystals (Fibers and Whiskers)

Cellulose materials are popular because of their availability, low density, renewability, high aspect ratio, high specific strength, modulus, and reactive surface [92]. Cellulose is mostly found in plants (fibers) and the cell wall of wood fibers as shown in Table 3.20. It consists of repeated crystalline structures with a collection of cellulose chains known as microfibrils. It is surrounded by an amorphous matrix of hemicelluloses and lignin substances. Amorphous regions act as structural defects and are responsible for the transverse cleavage of microfibrils into short monocrystals through acid hydrolysis. The methods used for nanofiber isolation are mechanical disintegration, acid hydrolysis, and biological treatments [130].

Table 3.20 Chemical components of wood fiber

Component	Soft wood	Hard wood
Cellulose	42 ± 2%	45 ± 2%
Hemicelluloses	27 ± 2%	30 ± 5%
Lignin	28 ± 2%	20 ± 4%
Extractives	3 ± 2%	5 ± 3%

Source: [162].

Cellulosic units consist of a primary wall, which is the first layer deposited during cell growth, and a secondary wall, which is composed of three layers. The thick middle layer consists of cellular microfibrils, and this layer decides the mechanical properties of the fiber. The angle between the fiber axis and microfibrils is known as microfibrillar angle [36, 106, 168]. These microfibrils have a diameter of about 10–30 nm and are made up of 30–100 cellulose molecules in extended chain conformation, which provides mechanical strength to the fiber [113, 153]. Cellulose without a degree of order is known as amorphous cellulose, which connects the elementary fibrils. This amorphous cellulose is also located between cellulose microfibrils. Cellulose properties with different engineering materials are shown in Table 3.21. Cellulose fibers on the nanoscale are developed by different methods: bacterial cellulose nanofibers, cellulose nanofibers by electro spinning, microfibrillated cellulose plant cell fibers, and nanorods or cellulose whiskers [5, 47].

Table 3.21 Properties of cellulose compared with different engineering materials

Material	Modulus (GPa)	Density (mg/m^3)	Specific modulus (GPa/mg m^3)
Crystalline cellulose	138	1.5	92
Aluminum	69	2.7	26
Steel	200	7.8	26
Glass	69	2.5	28

Source: [38].

Cellulose nanocrystals are also known as nanocrystalline cellulose or cellulose nanowhiskers or cellulose nanorods, which are elongated, single crystalline, rod-like particles of dimension of at least 1 nm. The width of cellulose nanocrystals is from 5 nm to 70 nm, and the length is from 100 nm to 250 nm (when plant cellulose is used as the source material) [113]. Cellulose nanocrystals are isolated from cellulose fibers through an acid treatment, which involves acid hydrolysis of biomass using concentrated sulfuric acid [170]. Microfibrillated cellulose consists of long, flexible, and entangled cellulose nanofibers prepared from different raw materials. Physical properties of cellulose fibril are shown in Table 3.22.

Table 3.22 Physical properties and sources of processed cellulose fibrils

Type	Origin of cellulose	Young's modulus (GPa)	Crystallinity (%)
Tunicate	Cellulose nanowhiskers	143	83–88
Cotton	Cellulose nanowhiskers	57–105	99
Bacterial cellulose	Glconacetobacter xylinus	114–138	89
Flax fiber	Microcrystalline cellulose	25	96

Source: [17].

Cellulose nanowhiskers are a fibrous form of cellulose produced by the acid hydrolysis of plant-based (bacterial/animal) cellulose, with lateral dimensions ranging from 3 nm to 30 nm and are used for nanocomposite applications. In the acid hydrolysis treatment process, amorphous regions of cellulose are dissolved, yielding crystalline rod/whisker shape nanoparticles with diameters ranging from 8 nm to 20 nm and lengths of 100 nm to a few micrometers. Whiskers have a high aspect ratio and high modulus, and their surface can be changed to various applications [75]. Cellulose nanowhiskers are not same as nanofibrillar cellulose (microfibrillated cellulose) [38]. Nanowhiskers give rise to strong and tough hydrogen-bonded networks, thereby improving the properties of composite materials, and are the most used cellulosic materials for reinforcing polymer matrices [17]. Cellulose nanoparticles with their sources are shown in Table 3.23.

Table 3.23 Different terminologies of cellulose nanoparticles and their sources with process

Name	Source	Process
Cellulose nanowhiskers	Ramie	H_2SO_4 hydrolysis
	Grass fiber	
	Cotton linters	HCl hydrolysis
Cellulose nanocrystals	Bacterial cellulose	H_2SO_4 hydrolysis
	Cotton	H_2SO_4 hydrolysis
Nanofibers	Wheat straw	HCl + mechanical treatment
Microfibrillated cellulose	Pulp Gaulin	Homogenizer
	Pulp Daicel	—
Nanofibrillated cellulose	Sulfite pulp	Mechanical
Microcrystalline cellulose	Alpha-cellulose fibers	Hydrolysis
Cellulose crystallites	Cotton Whatman filter paper	H_2SO_4 hydrolysis
Nanocellulose	Sisal fibers	H_2SO_4 hydrolysis
Nanofibers	Soybean pods	Chemical treatment + high pressure defibrillator

Source: [161].

Cellulose is the main constituent of plant structures, bacteria (acetobacter), and tunicates. Cellulose material can be obtained from results from different elements: biosynthesis of crystalline cellulose microfibrils (which is from cellulose source) and extraction process of cellulose particles (from cellulose microfibrils, which include treatment process). Tunicates (sea squirts) belong to a family of sea animals, and they are the only animals that produce cellulose microfibrils. They have a mantle consisting of cellulose microfibrils embedded in a protein matrix, and this thick leathery mantlein is used as a source of cellulose microfibrils [113]. Bacteria that produce cellulose are called *Gluconacetobacter xylinus* (from *Acetobacter xylinum*), and bacterial-derived cellulose microfibrils are possible to adjust culturing conditions to alter the formation of microfibril [14]. The species of algae (red, green, gray, yellow-green) produce cellulose microfibrils within the cell wall [113].

The two basic approaches for making nanostructures are bottom-up and top-down. The bottom-up approach involves producing molecular scale from scratch using atoms, molecules, and nanoparticles (method involving chemical synthesis and physics-derived technology). The top-down approach involves disintegration of macroscopic material to a nanoscale material by the following techniques: mechanical, chemical, enzymatic, and physical methods [8]. Properties of cellulose and other reinforcement materials are given in Table 3.24.

Table 3.24 Properties of cellulose and a few reinforcement materials

Material	ρ/g (per cm^3)	σ_f (GPa)	E_A (GPa)	E_T (GPa)
Crystalline cellulose	1.6	7.5–7.7	110–220	10–50
Kevlar fiber	1.4	3.5	124–130	2.5
Carbon fiber	1.8	1.5–5.5	150–500	—
Clay nanoplatelets	—	—	170	—
Boron nanowhiskers	—	2–8	250–360	—
Carbon nanotubes	—	11–63	270–950	0.8–30

Note: ρ = density, σ_f = tensile strength, E_A = elastic modulus in axial direction, E_T = elastic modulus in transverse direction [113].

Microcrystalline cellulose consists generally of stiff rod-like particles called whiskers. This microfibrillated cellulose is obtained through the mechanical treatment of pulp fibers, consisting of refining and high pressure homogenizing processes [73]. Cellulose nanocrystals extracted from cotton linter were milled to obtain fine particulate [142]. Studies on cellulose crystals prepared by acid hydrolysis of flax fiber were carried out [23]. It consists of slender rods with lengths ranging from 100 nm to 500 nm and diameters ranging from 10 nm to 30 nm. Nanoreinforced hydrogels prepared from wood cellulose whiskers are coated with chemically modified wood hemicelluloses [75], and water swollen hydrogels were investigated. Cellulose microfibril was prepared from banana rachis using a combination of chemical and mechanical treatments [184]. Bamboo cellulose nanocrystals were successfully prepared from the sulfuric acid hydrolysis of bamboo bleached fibers [21]. Cellulose microcrystals with dimensions of ~5 nm × 150–300 nm were obtained from wheat straw [55]. Nanocrystalline cellulose by controlled hydrolysis of bamboo fibers was also investigated [183].

The major drawback of cellulose whiskers with polar surfaces is poor compatibility with nonpolar solvents or resins. Cellulose nanoparticles and fibers are high in moisture absorption, incompatibility, limited processing temperature, and poor wetability. Plant cellulosic materials start to degrade approximately at 220 °C, restricting the use of natural fillers [101, 112, 161]. This can be overcome by surface modification such as surfactant coating or graft copolymerization [73, 83, 157, 181].

3.16 Conclusion

The use of natural or agricultural or cellulosic fibers as alternative fillers is increasing daily in today's world because of their initial low cost and availability. These fibers are favorable and are most needed from the ecology point of view. Wastewater treatment using cellulose fibers is new technology, and more research work is focused in implementing these fibers to the treatment. But every technology has its own pros and cons. Use of this technology is limited, and it depends on the type of application. Wastage of water should be minimized in all aspects, and if possible, recycled water should be used.

References

1. Abedi, M., and Bahreini, Z. (2010). Preparation of carbonaceous adsorbent from plant of *Calotropis gigantea* by thermochemical activation process and its adsorption behavior for removal of methylene blue, *World Appl. Sci. J.*, 11, pp. 263–268.
2. Abidin, A. S. B. Z. (2009). Adsorption of phenol by activated carbon produced from decanter cake, thesis report, Universiti Malaysia Pahang.
3. Alamri, H., and Low, I. M. (2013). Effect of water absorption on the mechanical properties of nano clay filled recycled cellulose fibre reinforced epoxy hybrid nanocomposites, *Compos A.*, 44, pp. 23–31.
4. Alawar, A., and Hamed, A. M. (2009). Characterization of treated date palm tree fiber as composite reinforcement, *Compos. B.*, 40, pp. 601–606.

5. Alemdar, A., and Sain, M. (2008). Isolation and characterization of nanofibres from agricultural residues – wheat straw and soy hulls. *Bioresour. Technol.*, 99, pp. 1664–1671.
6. Alsaeed, T., Yousif, B. F., and Ku, H., (2013). The potential of using date palm fibres as reinforcement for polymeric composites, *Mater. Des.*, 43, pp. 177–184.
7. Angelini, L. G., Lazzeri, A., Levita, G., Fontanelli, D., and Bozzi, C. (2000). Ramie (*Boehmeria nivea* L.) Gaud and Spanish Broom (*Spartium junceum* L.) fibers for composite materials: Agronomical aspects, morphology and mechanical properties. *Indl. Crops Prod.*, 11, pp. 145–161.
8. Antczak, M. S., Kazimierczak, J., and Antczak, T. (2012). Nanotechnology—Methods of manufacturing cellulose nanofibres, *Fibres Textl. Eastern Europe*, 2, pp. 8–12.
9. Babel, S., and Kurniawan, T. A. (2004). Cr (VI) removal from synthetic wastewater using coconut shell charcoal and commercial activated carbon modified with oxidizing agents and/or chitosan, *Chemosphere*, 54, pp. 951–967.
10. Baei, M. S., Babaee, V., and Pirouz, F. (2011). Preparation of poly aniline nanocomposites for removal of sulfate from wastewater, in *2nd International Conference on Chemistry and Chemical Engineering*, IACSIT Press, Singapore, pp. 95–100.
11. Baiardo, M., Zini, E., and Scandola, M. (2004). Flax fiber–polyester composites. *Compos. A.*, 35, pp. 703–710.
12. Bakar, B. F. A., Hiziroglu, S., and Tahir, P. M. (2013). Properties of some thermally modified wood species, *Mater. Des.*, 43, pp. 348–355.
13. Baley, C. (2002). Analysis of the flax fibers tensile behavior and analysis of the tensile stiffness increase. *Compos. A.*, 33, pp. 939–948.
14. Barud, H. S., Caiut, J. M. A., Ghys, J. D., Messaddeq, Y., and Ribeiro, S. J. L. (2012). Transparent bacterial cellulose boehmite epoxy siloxane nanocomposites, *Compos. A.*, 43, pp. 973–977.
15. Berger, F. T. W., and Weston, N. (2003). *Natural fibers, Plastics and Composites*, Kluwer Academic Publishers.
16. Berglund, L. (2005). Cellulose based nanocomposites, in *Natural fibres, Biopolymers, and Biocomposites* (Mohanty, A. K., Misra, M., Drzal, L. T., eds), CRC Press, Taylor & Francis Group, Boca Raton, FL, pp. 819–842.
17. Blaker, J. J., Lee, K. Y., and Bismarck, A. (2011). Hierarchical composites made entirely from renewable resources, *J. Biobased Mat. Bioenergy*, 5, pp. 1–16.

18. Bledzki, A., and Gassan, J. (1999). Composites reinforced with cellulose based fibers, *Prog. Poly. Sci.*, 24, pp. 221–274.
19. Bodzek, M., Konieczny, K., and Kwiecinska, A. (2011). Application of membrane processes in drinking water treatment–state of art, *Desalination Water Treat.*, 35, pp. 164–184.
20. Bos, H. L., Oever, M. J. A. V. D., and Peters, O. C. J. J. (2002). Tensile and compressive properties of flax fibers for natural fiber reinforced composites. *J. Mat. Sci.*, 37, pp. 1683–1692.
21. Brito, B. S. L., Pereira, F. V., Putaux, J. L., and Jean, B. (2012). Preparation, morphology and structure of cellulose nanocrystals from bamboo fibers, *Cellulose*, 19, pp. 1527–1536.
22. Callister, W. D. (1999). *Materials Science and Engineering: An Introduction*, John Wiley and Sons, New York.
23. Cao, X., Chen, Y., Chang, P. R., Muir, A. D., and Falk, G. (2008). Starch-based nanocomposites reinforced with flax cellulose nanocrystals, *eXPRESS Polym. Lett.*, 2, pp. 502–510.
24. Carey, R. O., and Migliaccio, K. W. (2009). Contribution of wastewater treatment plant effluents to nutrient dynamics in aquatic systems: A review, *Environ. Mgmt.*, 44, pp. 205–217.
25. Chandra, T. V. R., Ahalya, N., and Kanamadi, R. D., Biosorption: Techniques and Mechanisms, CES Technical Report 110, www.astra.iisc.ernet.in
26. Chakraborty, A., Deva, D., Sharma, A., and Verma, N. (2011). Adsorbents based on carbon microfibers and carbon nanofibers for the removal of phenol and lead from water, *J. Colloid. Interf. Sci.*, 359, pp. 228–239.
27. Cho, D., Lee, S. G., Park, W. H., and Han, S. O. (2002). Eco-friendly biocomposite materials using biofibers, *Polym. Sci. Technol.*, 13, pp. 460–476.
28. Chowdhury, Z. Z., Zain, S. M., Khan, R. A., and Ashraf, M. A. (2011). Preparation, characterization and adsorption performance of the KOH-activated carbons derived from kenaf fiber for lead (II) removal from wastewater, *Sci. Resrch. Essays*, 6, pp. 6185–6196.
29. Cichocki, Jr. F. R., and Thomason, J. L. (2002). Thermoelastic anisotropy of a natural fiber, *Compos. Sci. Technol.*, 62, pp. 669–678.
30. Crini, G. (2005). Recent developments in polysaccharide-based materials used as adsorbents in wastewater treatment, *Prog. Polym. Sci.*, 30, pp. 38–70.
31. Crini, G. (2006). Non-conventional low-cost adsorbents for dye removal: a review, *Bioresour. Techno.*, 97, pp. 1061–1085.

32. Das, P. K., Nag, D., Debanth, S., and Nayak, L. K. (2010). Machinery for extraction and traditional spinning of plant fibres, *Indian J. Trad. Knowldg.*, 9, pp. 386–393.
33. Din, A. T. M., Hameed, B. H., and Ahmad, A. L. (2009). Batch adsorption of phenol onto physiochemical-activated coconut shell, *J. Hazard. Mat.*, 161, pp. 1522–1529.
34. Doner, L. W., and Hicks, K. B. (1997). Isolation of hemicellulose from corn fiber by alkaline hydrogen peroxide extraction, *Cereal Chem.*, 74, pp. 176–181.
35. Donnel, L. A. O., Dweib, M. A., and Wool, R. P. (2004). Natural fiber composites with plant oil-based resin, *Compos. Sci. Technol.*, 64, pp. 1135–1145.
36. Eichhorn, S. J., Baillie, C. A., Zafeiropoulos, N., Mwaikambo, L. Y., Ansell, M. P., and Dufresne, A. (2001). Current international research into cellulosic fibers and composites, *J. Mater. Sci.*, 39, pp. 2107–2131.
37. Eichhorn, S. J., and Young, R. J. (2003). Deformation micromechanics of natural cellulose fibre networks and composites, *Compos. Sci. Technol.*, 63, pp. 1225–1230.
38. Eichhorn, S. J. (2011). Cellulose nanowhiskers: Promising materials for advanced applications, *Soft Matter.*, 7, pp. 303–315.
39. English, B., Youngquist, J. A., and Krzysik, A. M. (1994). Lignocellulosic composites, in *Cellulosic Polymers, Blends and Composites* (Gilbert, R. D., ed), Hanser Publishers, New York, pp. 115–130.
40. Espert, A., Vilaplana, F., and Karlsson, S. (2004). Comparison of water absorption in natural cellulosic fibres from wood and one-year crops in polypropylene composites and its influence on their mechanical properties, *Compos. A.*, 35, pp. 1267–1276.
41. Figueiredo, M. C. B., Rosa, M. F., Ugaya, C. M. L., Filho, M. S. M. S., Braid, A. C. C., and Melo, S. L. F. L. (2012). Life cycle assessment of cellulose nanowhiskers, *J. Cleaner Prodn.*, 35, pp. 130–139.
42. Fowler, P. A., Hughes, J. M., and Elias, R. M. (2006). Review biocomposites: Technology, environmental credentials and market forces, *J. Sci. Food. Agric.*, 86, pp. 1781–1789.
43. Franco, P. J. H., and Gonzalez, A. V. (2004). Mechanical properties of continuous natural fibre-reinforced polymer composites, *Compos. A.*, 35, pp. 339–345.
44. Fu, S. Y., and Lauke, B. (1996). Effects of fiber length and fiber orientation distributions on the tensile strength of short-fiber-reinforced polymers, *Compos. Sci. Technol.*, 56, pp. 1179–1190.

45. Fuadi, N. A. B., Ibrahim, A. S., and Ismail, K. N. (2012). Review study for activated carbon from palm shell used for treatment of wastewater, *J. Purity, Utility Reaction Environ.*, 1, pp. 252–266.
46. Fukazawa, K., Revol, J. F., Jurasek, L., and Goring, D. A. I. (1982). Relationship between ball milling and the susceptibility of wood to digestion by cellulose, *Wood Sci. Technol.*, 16, pp. 279–285.
47. Gardner, D. J., Oporto, G. S., Mills, R., and Samir, M. A. S. A. (2008). Adhesion and surface issues in cellulose and nanocellulose, *J. Adhesion Sci. Technol.*, 22, pp. 545–567.
48. Giancaspro, J., Papakonstantinou, C., and Balaguru, P. (2009). Mechanical behavior of fire-resistant biocomposite, *Compos. B.*, 40, pp. 206–211.
49. Gong, R., Cai, W., Li, N., Chen, J., Liang, J. and Cao, J. (2010). Preparation and application of thiol wheat straw as sorbent for removing mercury ion from aqueous solution, *Desalination Water Treat.*, 21, pp. 274–279.
50. Gounder, T. C., Shanker, M., and Nageswaran, S. (2011). Adsorptive removal of crystal violet dye using agricultural waste cocoa (*Theobroma cacao*) shell, *Res. J. Chem. Sci.*, 1, pp. 38–45.
51. Gutierrez, M. C., Paoli, M. A. D., and Felisberti, M. I. (2012). Biocomposites based on cellulose acetate and short curauá fibers: Effect of plasticizers and chemical treatments of the fibers, *Compos. A.*, 43, pp. 1338–1346.
52. Halpin, J. C., and Kardos J. L. (1978). Strength of discontinuous reinforced composites: I. Fiber reinforced composites, *Polym. Eng. Sci.*, 18, pp. 496–504.
53. Harish, S., Peter, M. D., Bensely, A., Mohan, L. D., and Rajadurai, A. (2009). Mechanical property evaluation of natural fiber coir composite, materials characterization, *Mater. Charact.*, 60, pp. 44–49.
54. Hashem, M. A. (2007). Adsorption of lead ions from aqueous solution by okra Wastes, *Intl. J. Phys. Sci.*, 2, pp. 178–184.
55. Helbert, W., Cavaille, J. Y., and Dufresne, A. (1996). Thermoplastic nanocomposites filled with wheat straw cellulose whiskers. Part I: processing and mechanical behavior, *Polym. Compo.*, 17, pp. 604–611.
56. Herrmann, A. S., Nickel, J., and Riedel, U. (1998). Construction materials based upon biologically renewable resources-from components to finished parts, *Polym. Degrad. Stab*, 59, pp. 251–261
57. Ho, M. P., Wang, H., Lee, J. H., Ho, C. K., Lau, K. T., Leng, J., and Hui, D. (2012). Critical factors on manufacturing processes of natural fibre composites, *Compos. B.*, 43, pp. 3549–3562.

58. Holbery, J., and Houston, D. (2006). Natural-fiber-reinforced polymer composites in automotive applications, *J. Manag.*, 58, pp. 80–86.
59. Hosseinaei, O., Wang, S., Enayati, A. A., and Rials, T. G. (2012). Effects of hemicellulose extraction on properties of wood flour and wood–plastic composites, *Compos. A.*, 43, pp. 686–694.
60. Hull, D., and Clyne, T. W. (1996). *An Introduction to Composite Materials*, Cambridge University Press, Cambridge, UK.
61. Husin, N. I., Wahab, N. A. A., Isa, N., and Boudville, R. (2011). Sorption equilibrium and kinetics of oil from aqueous solution using banana pseudostem fibers, in *International Conference on Environment and Industrial Innovation*, IACSIT Press, Singapore.
62. Husseien, M., Amer, A. A., Maghraby, A. E., and Taha, N. A. (2007). Utilization of barley straw as a source of a activated carbon for removal of methylene blue from aqueous solution, *J. Appl. Sci. Res.*, 3, pp. 1352–1358.
63. Ideriah, T. J. K., David, O. D., and Ogbonna, D. N. (2012). Removal of heavy metal ions in aqueous solutions using palm fruit fibre as adsorbent, *J. Environ. Chem. Eco. Toxicol.*, 4, pp. 82–90.
64. Igwe, J. C., and Abia, A. A. (2006). A bioseparation process for removing heavy metals from wastewater using biosorbents, *African J. Biotechnol.*, 5, pp. 1167–1179.
65. Ioannidou, O., and Zabaniotou, A. (2007). Agricultural residues as precursors for activated carbon production: A review, *Renew. Sustain. Energy Rev.*, 11, pp. 1966–2005.
66. Joffe, R., Andersons, J., and Wallstrom, L. (2003). Strength and adhesion characteristics of elementary flax fibers with different surface treatments, *Compos. A.*, 34, pp. 603–612.
67. John, M. J., and Thomas, S. (2008). Biofibers and biocomposites, *Carbohydr. Polym.*, 71, pp. 343–364.
68. Joseph, K., Filho, R. D, T., James, B., Thomas, S., and Carvalho, L. H. (1999). The use of sisal fiber as reinforcements in polymer in composites, *Brazilian J. Agri. Environ. Eng.*
69. Joshi, S. V., Drzal, L. T., Mohanty, A. K., and Arora, S. (2004). Are natural fiber composites environmentally superior to glass fiber reinforced composites, *Compos. A.*, 35, pp. 371–376.
70. Julian, F. V., and Vincent, V. (2000). A unified nomenclature for plant fibers for industrial use, *Appl. Compos. Mat.*, 7, pp. 269–271.
71. Juntaro, J., Ummartyotin, S., Sain, M., and Manuspiy, H. (2012). Bacterial cellulose reinforced polyurethane-based resin nanocomposite: A study

of how ethanol and processing pressure affect physical, mechanical and dielectric properties, *Carbohydr. Polym.*, 87, pp. 2464–2469.

72. Kalia, S., Dufresne, A., Cherian, B. M., Kaith, B. S., Averous, L., Njuguna, J., and Nassiopoulos, E. (2011). Cellulose-based bio- and nanocomposites: A review, *Intl. J. Polym. Sci.*, doi:10.1155/2011/837875.

73. Kamel, S. (2007). Nanotechnology and its applications in lingo cellulosic composites, a mini review, *eXPRESS Polym. Lett.*, 1, pp. 546–575.

74. Kanawade, S. M., Gaikwad, R. W., and Misal, S. A. (2010). Low cost sugarcane bagasse ash as an adsorbent for dye removal from dye effluent, *Intl. J. Chem. Eng. Appl.*, 1, pp. 309–318.

75. Karaaslan, M. A., Tshabalala, M. A., Yelle, D. J., and Diller, G. B. (2011). Nano-reinforced biocompatible hydrogels from wood hemicelluloses and cellulose whiskers, *Carbohydr. Polym.*, 86, pp. 192–201.

76. Kelly, A., and Tyson, W. R. (1965). Tensile properties of fiber-reinforced metals: Copper/tungsten and copper/molybdenum, *J. Mech. Phys. Solids*, 13, pp. 329–350.

77. Kestur, S. G., Sahagun, T. H. S. F., Santos, L. P. D., Santos, J. D., Mazzaro, I., and Mikowski, A. (2013). Characterization of blue agave bagasse fibers of Mexico, *Compos. A.*, 45, pp. 153–161.

78. Klasson, K. T., Wartelle, L. H., Lima, I. M., Marshall, W. E., and Akin, D. E. (2009). Activated carbons from flax shive and cotton gin waste as environmental adsorbents for the chlorinated hydrocarbon trichloroethylene, *Bioresour. Technol.*, 100, pp. 5045–5050.

79. Kowalczyk, M., Piorkowska, E., Kulpinski, P., and Pracella, M. (2011). Mechanical and thermal properties of PLA composites with cellulose nanofibers and standard size fibers, *Compos. A.*, pp. 1509–1514.

80. Krishnan, K. B, Doraiswamy, I., and Chellamani, K. P. (2005). Jute, in *Bast and other Plant Fibres* (Franck R. R., ed), Wood Head Publishing, Cambridge.

81. Ku, H., Wang, H., Pattarachaiyakoop, N., and Trada, M. (2011). A review on the tensile properties of natural fiber reinforced polymer composites, *Compos. B.*, 42, pp. 856–873.

82. Kumaar, S. S., Krishna, S. K., Kalaamani, P., Subburamaan, C. V., and Subramaniam, N. G. (2010). Adsorption of organophosphorous pesticide from aqueous solution using waste jute fiber carbon, *Modern Appl. Sci.*, 4, pp. 67–03.

83. Kumar, M. B., Pavithran, C., and Pillai, R. M. (2005). Coconut fibre reinforced polyethylene composites: Effect of natural waxy surface

layer of the fibre on fibre/matrix interfacial bonding and strength of composites, *Compos. Sci. Technol.*, 65, pp. 563–569.

84. Kumar, P., Agnihotri, R., Wasewar, K. L., Uslu, H., and Yoo, C. K. (2012). Status of adsorptive removal of dye from textile industry effluent, *Desalination Water Treat.*, 50, pp. 226–244.

85. Kuranska, M., and Prociak, A. (2012). Porous polyurethane composites with natural fibres, *Compos. Sci. Technol.*, 72, pp. 299–304.

86. Ladhe, U. V., Wankhede, S. K., Patil, V. T., and Patil, P. R. (2011). Adsorption of erichrome black T from aqueous solutions on activated carbon prepared from mosambi peel, *J. Appl. Sci. Environ. Sanit.*, 6, pp. 149–154.

87. Lam, S. M., Sin, J. C., Abdullah, A. Z., and Mohamed, A. R. (2012). Degradation of wastewaters containing organic dyes photo catalysed by zinc oxide: A review, *Desalination Water Treat.*, 41, pp. 131–169.

88. Latif, M. M. A. E., and Ibrahim, A. M. (2009). Adsorption, kinetic and equilibrium studies on removal of basic dye from aqueous solutions using hydrolyzed oak sawdust, *Desalination Water Treat.*, 6, pp. 252–268.

89. Lauke, B., and Fu, S. Y. (1999). Strength anisotropy of misaligned short-fiber reinforced polymers, *Compos. Sci. Technol.*, 59, pp. 699–708.

90. Lavoine, N., Desloges, I., Dufresne, A., and Bras, J. (2009). Microfibrillated cellulose—Its barrier properties and applications in cellulosic materials: A review, *Carbohydr. Polym.*, 78, pp. 422–431.

91. Lee, K. Y., Bharadia, P., Blaker, J. J., and Bismarck, A. (2012). Short sisal fibre reinforced bacterial cellulose poly lactide nanocomposites using hairy sisal fibres as reinforcement, *Compos. A.*, 43, pp. 2065–2074.

92. Li, Y., and Ragauskas, A. J. (2011). Cellulose nanowhiskers as a reinforcing filler in polyurethanes, in *Advances in Diverse Industrial Applications of Nanocomposites* (Reddy, B., ed), InTech, pp. 17–36.

93. Li, Y., Mai, Y. W., and Ye, L. (2000). Sisal and its composites: A review of recent developments, *Compos. Sci. Technol.*, 60, pp. 2037–2055.

94. Lilholt, H., and Lawther, J. M. (2000). Natural organic fibers, in *Comprehensive Composite Materials: Fiber Reinforcements and General Theory of Composites* (Chou, T. W., ed), Elsevier, New York, pp. 303–325.

95. Liu, D., Song, J., Anderson, D. P., Chang, P. R., and Hua, Y. (2012). Bamboo fiber and its reinforced composites: Structure and properties, *Cellulose*, 19, pp. 1449–1480.

96. Lokman, F. B. (2006). Dye removal from simulated wastewater by using empty fruit bunch as an adsorbent agent, Thesis Report, Universiti Malaysia Pahang.
97. Lua, A. C., and Guo, J. (2001). Adsorption of sulfur dioxide on activated carbon from oil-palm waste, *J. Environ. Eng.*, 127, pp. 895–901.
98. Luiz, C. A., Oliveiraa, Riosa, R. V. R. A., Fabrisa, J. D., Gargc, V., Sapagb, K., and Lagoa, R. M. (2002). Activated carbon/iron oxide magnetic composites for the adsorption of contaminants in water, *Carbon*, 40, pp. 2177–2183.
99. Macedo, J. S., Junior, N. B. C., Almeida, L. E., Vieira, E. F. S., Cestari, A. R., Gimenez, I. F., Carreno, N. L. V., and Barreto, L. S. (2006). Kinetic and calorimetric study of the adsorption of dyes on mesoporous activated carbon prepared from coconut coir dust, *J. Colloid. Inter. Sci.*, 298, pp. 515–522.
100. Madsen, B., Hoffmeyer, P., and Lilholt, H. (2012). Hemp yarn reinforced composites—III. Moisture content and dimensional changes, *Compos. A.*, 43, pp. 2151–2160.
101. Manfredi, L. B., Rodrıguez, E. S., Przybylak, M. W., and Zquez, A. V. (2006). Thermal degradation and fire resistance of unsaturated polyester modified acrylic resins and their composites with natural fibres, *Polym. Degrad. Stab.*, 91, pp. 255–261.
102. Marsh, G. (2003). Next step for automotive materials, *Mat. Today*, 6, pp. 36–43.
103. Matthews, F. L., and Rawlings, R. D. (1994). *Composite Materials: Engineering and Science,* Chapman & Hall, London.
104. Mohan, D., and Singh, K. P. (2002). Single and multi-component adsorption of cadmium and zinc using activated carbon derived from bagasse an agricultural waste, *Water Res.*, 36, pp. 2304–2318.
105. Mohan, T. P., and Kanny, K. (2012). Chemical treatment of sisal fiber using alkali and clay method, *Compos. A.*, 43, pp. 1989–1998.
106. Mohanty, A. K., Misra, M., and Hinrichsen, G. (2001). Biofibers, biodegradable polymers and biocomposites: An overview, *Macromol. Mat. Engg.*, 276–277, pp. 1–24.
107. Mohanty, A. K., Liu, W., Tummala, P., Drzal, L. T., and Misra, M. (2005). Soy protein based plastics, blends and composites, in *Natural Fibres, Biopolymers, and Biocomposites* (Mohanty, A. K., Misra, M., and Drzal, L. T., eds), CRC Press, Taylor & Francis Group, Boca Raton, FL, pp. 699–725.

108. Mohanty, A. K., Misra, M., and Drzal, L. T. (2001). Surface modifications of natural fibers and performance of the resulting bio composites: An overview, *Compos. Interf.*, 8, pp. 313–343.

109. Mohanty, A. K., Misra, M., and Drzal, L. T. (2002). Sustainable biocomposites from renewable resources: Opportunities and challenges in the green materials world, *J. Polym. Environ.*, 10, pp. 19–26.

110. Mohanty, A. K., Misra, M., Drzal, L. T., Selke, S. E., Harte, B. R., and Hinrichsen, G. (2005). Natural fibres, biopolymers, and biocomposites: An introduction, in *Natural Fibres, Biopolymers, and Biocomposites* (Mohanty, A. K., Misra, M., and Drzal, L. T., eds), CRC Press, Taylor & Francis Group, Boca Raton, FL, pp. 1–35.

111. Mohanty, K., Drzal, L. T., Hokens, D., and Misra, M. (2001). Eco-friendly composite materials from biodegradable polymers: Biocomposites to nanocomposites, polymer, *Polym. Mat. Sci. Eng.*, 85, 594.

112. Monteiro, S. N., Calado, V., Rodriguez, R. J. S., and Margem, F. M. (2012). Thermo gravimetric behavior of natural fibers reinforced polymer composites: An overview, *Mater. Sci. Eng.*, 557, pp. 17–28.

113. Moon, R. J., Martini, A., Nairn, J., Simonsen, J., and Youngblood, J. (2011). Cellulose nanomaterials review: Structure, properties and nanocomposites, *Chem. Soc. Rev.*, 40, pp. 3941–3994.

114. Mundhe, K. S., Gaikwad, A. B., Kale, A. A., Deshpande, N. R., and Kashalkar, R. V. (2012). Polyalthia longifolia (Ashoka) seeds: An effective adsorbent for methylene blue removal, *Intl. J. Res. Pharma. Biomed. Sci.*, 3, pp. 180–186.

115. Murugan, P. V., Kumar, V. R., and Karan, G. D. (2011). Dye removal from aqueous solution using low cost adsorbent, *Intl. J. Environ. Sci.*, 1, 7, pp. 1493–1503.

116. Musapatika, E. T., Singh, R., Moodley, K., Nzila, C., Onyango, M. S., and Ochieng, A. (2012). Cobalt removal from wastewater using pine sawdust, *African J. Bio. Technol.*, 11, pp. 9407–9415.

117. Mwaikambo, L. Y. (2006). Review of the history, properties and application of plant fibres, *African J. Sci. Tech.*, 7, pp. 120–133.

118. Nakanishi, S. C., Gonçalves, A. R., Rocha, G., Ballinas, M. L., and Gonzalez, G. (2011). Obtaining polymeric composite membranes from lingo cellulosic components of sugarcane bagasse for use in wastewater treatment, *Desalination Water Treat.*, 27, pp. 66–71.

119. Nascimento, D. C. O., Ferreira, A. S., Monteiro, S. N., Aquino, R. C. M. P., and Kestur, S. G. (2012). Studies on the characterization of piassava fibers and their epoxy composites, *Compos. A.*, 43, pp. 353–362.

120. Nascimento, R. F., Sousa, F. W., Neto, V. O. S., Fechine, P. B. A., Teixeira, R. N. P., Freire, P. T. C., and Silva, M. A. A. (2012). Biomass adsorbent for removal of toxic metal ions from electro plating industry waste, *Electroplating* (Sebayang, D., ed), Intech, pp. 101–136.

121. Nechwatal, A., Mieck, K. P., and Reubmann, T. (2003). Developments in the characterization of natural fibre properties and in the use of natural fibres for composites, *Compos. Sci. Technol.*, 63, pp. 1273–1279.

122. Nemeth, D., Labidi, J., Gubicza, L., and Bako, K. B. (2011). Comparative study on heavy metal removal from industrial effluents by various separation methods, *Desalination Water Treat.*, 35, pp. 242–246.

123. Ngaha, W. S. W., Teonga, L. C., and Hanafiaha, M. A. K. M. (2011). Adsorption of dyes and heavy metal ions by chitosan composites: A review, *Carbohydr. Polym.*, 83, pp. 1446–1456.

124. Nishino, T. (2004). Natural fibre sources, *Green Composites: Polymer Composites and the Environnent* (Baillie, C., ed), Wood Head Publishing, Cambridge, pp. 49–80.

125. Okoronkwo, A. E., and Olusegun, S. J. (2012). Biosorption of nickel using unmodified and modified lignin extracted from agricultural waste, *Desalination Water Treat.*, DOI: 10.1080/19443994. 2012.714896.

126. Okoye, I., Ejikeme, P. M., and Onukwuli, O. D. (2010). Lead removal from wastewater using fluted pumpkin seed shell activated carbon: Adsorption modeling and kinetics, *Int. J. Environ. Sci. Technol.*, 7, pp. 793–800.

127. Oksman, K., Skrifvars, M., and Selin, J. F. (2003). Natural fibers as reinforcement in poly lactic acid (PLA) composites, *Compos. Sci. Technol.*, 63, pp. 1317–1324.

128. Oliveira, E. A., Montanher, S. F., and Rollemberg, M. C. (2011). Removal of textile dyes by sorption on low-cost sorbents. A case study: Sorption of reactive dyes onto Luffa cylindrical, *Desalination Water Treat.*, 25, pp. 54–64.

129. Owabor, C. N., Agarry, S. E., and Jato, D. (2012). Removal of naphthalene from aqueous system using unripe orange peel as adsorbent: effects of operating variables, *Desalination Water Treat.*, 48, pp. 315–319.

130. Panaitescu, D. M., Frone, A. N., Ghiurea, M., Spataru, C. I., Radovici, C., and Iorga, M. D. (2011). Properties of polymer composites with cellulose microfibrils, *Adv. Compos. Mater. Eco. Des. Analy.*, pp. 103–122.

131. Pang, S. C., Chin, S. F., and Yih, V. (2011). Conversion of cellulosic waste materials into nanostructured ceramics and nanocomposites, *Adv. Mat. Lett.*, 2, pp. 118–124.

132. Panthapulakkal, S., and Sain, M. (2012). Preparation and characterization of cellulose nanofibril films from wood fibre and their thermoplastic polycarbonate composites, *Intl. J. Polym. Sci.*, Article ID 381342, 6 pages.
133. Peng, B. L., Dhar, N., Liu, H. L., and Tam, K. C. (2011). Chemistry and applications of nanocrystalline cellulose and its derivatives: A nanotechnology perspective, *Can. J. Chem. Eng.*, 9999, pp. 1–16.
134. Phuong, V. T., and Lazzeri, A. (2012). Green biocomposites based on cellulose diacetate and regenerated cellulose microfibers: Effect of plasticizer content on morphology and mechanical properties, *Compos. A.*, 43, pp. 2256–2268.
135. Placet, V., Trivaudey, F., Cisse, O., Retel, V. G., and Boub, M. L. (2012). Diameter dependence of the apparent tensile modulus of hemp fibres: A morphological, structural or ultrastructural effect. *Compos. A.*, 43, pp. 275–287.
136. Pothan, L. A., and Thomas, S. (2004). Effect of hybridization and chemical modification on the water-absorption behavior of banana fiber reinforced polyester composites, *J. Appl. Polym. Sci.*, 91, pp. 3856–3865.
137. Poura, A. M., Zabihia, M., Tahmasbib, M., and Bastamib, T. R. (2010). Effect of adsorbents and chemical treatments on the removal of strontium from aqueous solutions, *J. Hazard. Mater.*, 182, pp. 552–556.
138. Prasad, B. M., Sain, M. M., and Roy, D. N. (2005). Properties of ball milled thermally treated hemp fibres in an inert atmosphere for potential composite reinforcement, *J. Mat. Sci.*, 40, pp. 4271–4278.
139. Prasad, J., and Gupta, C. B. (1975). Mechanical properties of maize stalks as related to harvesting, *J. Agri. Eng. Res.*, 20, pp. 79–87.
140. Quan, Y. S., Yuan, G. S., Gang, Y. Y., Hui, W., and Rui, H. (2012). Removal of the heavy metal ion Cr(VI) by soybean hulls in dye house wastewater treatment, *Desalination Water Treat.*, 42, pp. 197–201
141. Rajan, M. J., Arunachalam, R., and Durai, G. A. (2011). Agricultural wastes of jackfruit peel nano-porous, adsorbent for removal of Rhodamine dye, *Asian J. Appl. Sci.*, 4, pp. 263–270.
142. Roohani, M., Habibi, Y., Belgacem, N. M., Ebrahim, G., Karimi, A. N., and Dufresne, A. (2008). Cellulose whiskers reinforced polyvinyl alcohol copolymers nanocomposites, *Eur. Polym. J.*, 44, pp. 2489–2498.
143. Rowell, R. M. (2002). *Sustainable Composites from Natural Resources from High Performance Structures and Composites* (Brebbia, C. A., and Wilde, W. P. D., eds), WIT Press, Southampton, Boston, pp. 183–192.

144. Rowell, R. M., and Stout, H. P. (1998). Jute and kenaf, in *Handbook of Fibre Chemistry* (Lewis, M., and Pearce, E. M., eds), 2nd Edition, CRC Press, New York, pp. 465–504.

145. Rowell, R. M., Sanadi, A. R., Caulfield, D. F., and Jacobson, R. E. (1997). Utilization of natural fibres in plastic composites: Problems and opportunities, *Proc. Intl. Symp. Lignocellulosic-Plastic Comp*, Sao Paulo, pp. 23–51.

146. Rubio, J., Souza, M. L., and Smith, R. W. (2002). Overview of flotation as a wastewater treatment technique, *Minerals Eng.*, 15, pp. 139–155.

147. Saheb, D. N., and Jog, J. P. (1999). Natural fiber polymer composites: A review, *Adv. Polym. Technol.*, 18, pp. 351–363.

148. Saksule, A. S., and Kude, P. A. (2012). Adsorbents from karanja seed oil cake and applications, *Intl. J. Chem. Eng. Appl. Sci.*, 2, pp. 13–23.

149. Sakthi, V., Andal, N. M., Rengaraj, S., and Sillanpaa, M. (2010). Removal of Pb (II) ions from aqueous solutions using *Bombax ceiba* sawdust activated carbon, *Desalination Water Treat.*, 16, pp. 262–270.

150. Salemane, M. G., and Luyt, A. S. (2006). Thermal and mechanical properties of polypropylene wood powder composites, *J. Appl. Polym. Sci.*, 100, pp. 4173–4180.

151. Salvado, A. P. A., Campanholi, L. B., Fonseca, J. M., Tarley, C. R. T., Caetano, J., and Dragunski, D. C. (2012). Lead (II) adsorption by peach palm waste, *Desalination Water Treat.*, 48, pp. 335–343.

152. Sarin, V., and Pant, K. K. (2006). Removal of chromium from industrial waste by using eucalyptus bark, *Bioresour. Technol.*, 97, pp. 15–20.

153. Satyanarayana, K. G. (2007). Biodegradable composites based on lignocellulosic fibers, International Conference on Advanced Materials and Composites.

154. Sauepraseasrit, P., Nuanjaraen, M., and Chinlapa, M. (2010). Biosorption of lead (Pb^{2+}) by luffa cylindrical fiber, *Environ. Res. J.*, 4, pp. 157–166.

155. Schemanauer, J. J., Osswald, T. A., Sanadi, A. R., and Caulfield, D. F. (2000). Melt rheological properties of natural fiber-reinforced polypropylene. *Proc. ANTEC*, pp. 2206–2210.

156. Scott, G. M., and Smith, A. (1995). Sludge characteristics and disposal alternatives for recycled fiber plants, *Recy. Symp.*, pp. 239–249.

157. Sgriccia, N., Hawley, M. C., and Misra, M. (2008). Characterization of natural fiber surfaces and natural fiber composites, *Compos. A.*, 39, pp. 1632–1637.

158. Shareef, K. M. (2009). Sorbents for contaminants uptake from aqueous solutions. Part I—Heavy metals, *World J. Agril. Sci.*, 5 (S), pp. 819–831.

159. Sharma, H. S. S. (1987). Studies on chemical and enzyme retting of flax on a semi-industrial scale and analysis of the effluents for their physicochemical components, *Intl. Biodeter.*, 3, pp. 329–342.

160. Shekeil, Y. A. E., Sapuan, S. M., Khalina, A., Zainudin, E. S., and Al-Shuja, O. M. (2012). Influence of chemical treatment on the tensile properties of kenaf fiber reinforced thermoplastic polyurethane composite, *eXPRESS Polym. Lett.*, 6, pp. 1032–1040.

161. Siqueira, G., Bras, J., and Dufresne, A. (2010). Cellulosic bionanocomposites: A review of preparation, properties and applications, *Polymer*, 2, pp. 728–765.

162. Smook, G. A. (1999). *Handbook for Pulp and Paper Technologists*, 2nd edition, Angus Wilde Publications Inc, Vancouver, Canada.

163. Spence, K. L., Venditti, R. A., Habibi, Y., Rojas, O. J., and Pawlak, J. J. (2010). The effect of chemical composition on microfibrillar cellulose films from wood pulps: Mechanical processing and physical properties, *Biores. Technol.*, 101, pp. 5961–5968.

164. Sreekala, M. S., Kumaran, M. G., and Thomas, S. (2002). Water sorption in oil palm fiber reinforced phenol formaldehyde composites. *Compos. A.*, 33, pp. 763–777.

165. Sugumaran, P., Susan, V. P., Chandran, P. R., and Seshadri, S. (2012). Production and characterization of activated carbon from banana empty fruit bunch and *Delonix regia* fruit pod, *J. Sustain. Energ. Environ.*, 3, pp. 125–132.

166. Sulak, M. T., and Yatmaz, H. C. (2012). Removal of textile dyes from aqueous solutions with eco-friendly biosorbent, *Desalination Water Treat.*, 37, pp. 169–177.

167. Tan, I. A. W., Ahmad, A. L., and Hameed, B. H. (2008). Preparation of activated carbon from coconut husk: Optimization study on removal of 2,4,6-trichlorophenol using response surface methodology, *J. Hazard. Mat.*, 153, pp. 709–717.

168. Thomas, S., Paul, S. A., Pothan, L. A., and Deepa, B. (2011). Natural fibers: Structure, properties and applications, in *Cellulose Fibers: Bio- and Nano-Polymer Composites* (Kalia, S., Kaith, B. S., and Kaur, I., eds), Springer-Verlag, New York, pp. 3–42.

169. Thumm, A., and Dickson, A. R. (2013). The influence of fiber length and damage on the mechanical performance of polypropylene/wood pulp composites, *Compos. A.*, 46, pp. 45–52.

170. Tingaut, P., Eyholzer, C., and Zimmermann, T. (2011). Functional polymer nanocomposite materials from microfibrillated cellulose, in

Advances in Nanocomposite Technology (Hashim, A., ed), Intech, pp. 319–334.

171. Tsai, W. T., Chang, C. Y., Lin, M. C., Chein, S. F., Sun, H. F., and Hsieh, M. F. (2001). Adsorption of acid dye onto activated carbons prepared from agricultural waste bagasse by $ZnCl_2$ activation, *Chemosphere*, 45, pp. 51–58.

172. Tseng, R. L., Wu, F. C., and Juang, R. S. (2003). Liquid-phase adsorption of dyes and phenols using pinewood-based activated carbons, *Carbon*, 41, pp. 487–495.

173. Urkiaga, A., and Fuentes, L. D. L. (2006). Best available technologies for water reuse and recycling needed steps to obtain the general implementation of water reuse.

174. Urruzola, I., Andres, M. A., Nemeth, D., Bako, K. B., and Labidi, J. (2012). Multicomponents adsorption of modified cellulose microfibrils, *Desalination Water Treat.*, 51, pp. 2153–2161.

175. Velde, K. V. D., and Kiekens, P. (2002). Biopolymers: Overview of several properties and consequences on their applications, *Polym. Test.*, 21, pp. 433–442.

176. Velde, K. V. D., and Kiekens, P. (2001). Thermoplastic polymers: Overview of several properties and their consequences in flax fiber reinforced composites. *Polym. Test.*, 20, pp. 885–893.

177. Visakh, P. M., Thomas, S., Oksman, K., and Mathew, A. P. (2012). Effect of cellulose nanofibers isolated from bamboo pulp residue on vulcanized natural rubber, *Bioresour.*, 7, pp. 2156–2168.

178. Wang, P., Zhang, R., and Hua, C. (2012). Removal of chromium (VI) from aqueous solutions using activated carbon prepared from crofton weed, *Desalination Water Treat.*, 51, 2327–2335.

179. Wang, W., Sain, M., and Cooper, P. A. (2006). Study of moisture absorption in natural fiber plastic composites, *Compos. Sci. Technol.*, 66, pp. 379–386.

180. Wool, R., and Sun, X. S. (2005). *Bio-Based Polymers and Composites*, Elsevier, Burlington, MA.

181. Xie, Y., Hill, C. A. S., Xiao, Z., Militz, H., and Mai, C. (2010). Silane coupling agents used for natural fiber/polymer composites: A review, *Compos. A.*, 41, pp. 806–819.

182. Young, R. A., and Rowell, R. M. (1986). *Cellulose: Structure, Modification and Hydrolysis*, John Wiley & Sons, New York.

183. Zhang, Y., Lu, X. B., Gao, C., Lv, W. J., and Yao, J. M. (2012). Preparation and characterization of nanocrystalline cellulose from bamboo fibers

by controlled cellulase hydrolysis, *J. Fiber Bio. Eng. Inform.,* 5, pp. 263–271.

184. Zuluaga, R., Putaux, J. L., Restrepo, A., Mondragon, I., and Ganan, P. (2007). Cellulose microfibrils from banana farming residues: Isolation and characterization, *Cellulose,* 14, pp. 585–592.

Chapter 4

Removal of Heavy Metals from Water Using PCL, EVA–Bentonite Nanocomposites

Derrick S. Dlamini, Ajay K. Mishra, and Bhekie B. Mamba
Department of Chemical Technology, University of Johannesburg, P.O. Box 17011, Doornfontein, Corner Nind and Beit street, Johannesburg 2028, South Africa
derricksdlamini@gmail.com

4.1 Introduction

Over the past few decades, pollution of water resources due to disposal of heavy metals, especially from industrial activities, has been causing worldwide concern. Unlike organic water pollutants, which are generally susceptible to biodegradation and photodegradation, heavy metals cannot be destroyed. Heavy metals are a great threat to aquatic life even at relatively low concentrations. Some heavy metals are capable of being assimilated, stored, and concentrated in human bodies, causing erythrocyte destruction, nausea, salivation, muscular cramps, renal degradation, chronic pulmonary problems, and skeletal deformity. Transition metals such as arsenic (As), chromium

(Cr), copper (Cu), lead (Pb), and mercury (Hg) are considered to be toxic to human and aquatic life.

Several methods have been employed for the removal of heavy metals in water. The widely used methods for heavy-metal uptake from aqueous solutions include chemical precipitation, membrane filtration, ion exchange, and adsorption among others. Adsorption is one of the most effective and simplest approaches for the removal of toxic and recalcitrant pollutants from aqueous systems. Several adsorbents for removal of arsenic, cadmium, and lead from contaminated waters have been reported. Adsorption is the most commonly used technique because potentially there are several materials that can function as adsorbents for heavy metals. Another adsorbent slowly gaining popularity in water treatment is polymer–clay nanocomposite (PCN). This chapter looks into the possibility of using PCNs in wastewater treatment specifically for the removal of heavy metals.

4.2 Polymeric Nanocomposites

Polymeric composites are polymers that have been filled with synthetic or natural inorganic compounds to improve their chemical and physical properties or to reduce cost by acting as a diluent for the polymer [1–3]. In this context, the polymer is called a matrix. If the filler is in the nanometer range, the composite is called a nanocomposite. These nanocomposites are a new class of composites, and they possess unique properties, which are typically not shared by their more conventional microscopic counterparts [2]. The unique properties are attributed to the nanometer size features and the extraordinarily high surface area of the dispersed nanoparticles [2, 3], and the interaction between the polymer and the clay particles. Many fillers have been used in nanocomposites preparation, but clay (hydrous layer silicates) and layered silicates have been most widely used [4–6]. Clay-based nanocomposites, which are the subject of this research, have received considerable scientific and technological attention due to the fact that clay is environmentally friendly, abundantly available, and there is wide knowledge available on clay-intercalation chemistry [7]. Clays consist mainly of clay minerals and carbonates, subordinate quartz, anatase, feldspars, and goethite, with the last two being present only in the calcite-rich clay as trace constituents [1]. According to the characteristics of layer type (1:1, or 2:1), the magnitude of net layer charge (x) per formula unit, and the type of interlayer species, clay minerals can be classified into seven groups: (1) kaolin-serpentine, (2) pyrophyllite-talc, (3) smectite, (4) vermiculite, (5) mica, (6) chlorite, and (7) interstratified clay minerals [2]. Smectite clays are very useful for many applications in the field of catalysis, ion exchange, and adsorption [3]. Smectite clays belong to a family of layered minerals that consist of individual platelets with a metal oxide center sandwiched between two silicone dioxide outer layers. Included in this group of minerals are hectorite, bentonite, montmorillonite (MMT), saponite, sepiolite, beidellite, nontronite, and sauconite. Of these, hectorite, MMT, and bentonite are the most important because of their swelling properties and availability.

4.2.1 Nanocomposite Formation and Structure

4.2.1.1 Polymer–clay nanocomposite formation

Any physical mixture of a polymer and clay does not essentially form a nanocomposite with improved properties. Immiscibility leads to poor physical attraction between the polymer and the clay particles and that consequently leads to relatively poor mechanical properties. Generally, clays are hydrophilic, which means it is necessary to select a hydrophilic polymer to avoid immiscibility when preparing nanocomposites. However, hydrophilic polymers are not compatible with all composite-fabrication procedures. Fortunately, the clay can be modified accordingly to enhance compatibility. In many cases, a surfactant is used and the resulting modified clay is called organoclay. Many compounds have been used to modify clay. For example, dimethyl sulfoxide, methanol, and octadecylamine were used by Cabedo et al. [8] to increase the basal plane distance of the original clay by a factor of more than three. The "popular" Cloisite® organoclays are prepared by modifying MMT clay with salts of dimethyl dehydrogenated tallow quaternary ammonium.

The concept of modifying the clay basically takes advantage of the exchangeable ions in the gallery of the clay. The general structure of a 2:1 dioctahedral smectite clay is shown in Fig. 4.1.

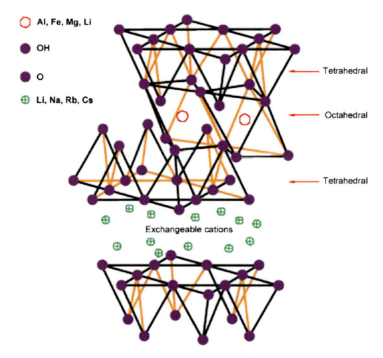

Figure 4.1   Layered silicate structure; adapted from Beyer [9].

The structure shows that the clay possesses Li, Na, Rb, and Cs ions sandwiched between two layers. These ions can be replaced with surfactants under appropriate conditions to form an organoclay. The replacement of the exchange cations in the cavities of the layered clay structure by alkylammonium surfactants can compatibilize the surface chemistry of the clay and a hydrophobic polymer matrix. Madaleno et al. [10] reported that when natural MMT is used, the dispersion of the clay in the polyvinyl chloride (PVC) matrix is typically poor, and when organically modified MMT is used, an improved dispersion is observed. Therefore, organic modification of the clay is very important, and it has an effect on the nanocomposite properties. It should be mentioned that the type of surfactant, the chain length, and the packing density may play an important role [11] in determining the suitability of the surfactant to modify the clay. Replacing the exchangeable ions in the cavity of the clay does not only serve to alter the hydrophilic or polarity properties of the clay, but also widens the inter-gallery spacing (illustrated in Fig. 4.2) to allow unhindered intercalation of the polymer chains in the clay. By definition, intercalation is the insertion of polymer chains between two galleries of clay. This enlarges the inter-gallery spacing of the clay.

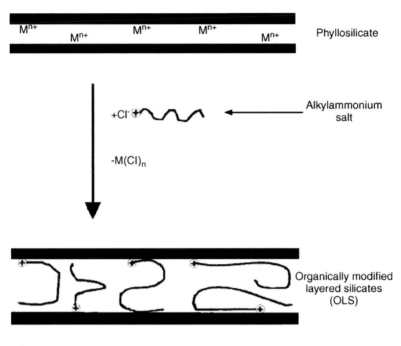

M^{n+} Metal cation

 Alkyl ammonium cation

Figure 4.2   Schematic illustration of the modification of clay; adapted from Zanetti et al. [12].

The expansion of the gallery spacing by the organic modifier allows proper formation of a nanocomposite during fabrication. Three strategies are mainly employed in the preparation of nanocomposites: in situ polymerization [13], melt-blending method, and solution-blending method [10, 14].

4.2.1.1.1 *In situ polymerization*

In the in situ polymerization procedure, monomers are mixed with the clay prior to polymerization. The main goal is to disperse the clay layers in the polymer matrix to obtain a nanocomposite with homogeneous exfoliated structures. Uniform dispersion of the clay in this method is attributed to the low monomer viscosity. According to Lingaraju et al. [15], this procedure was used in the preparation of the "first" nanocomposite, whereby researchers at the Toyota Central Research Laboratories synthesized nylon 6–clay nanocomposite.

For most technologically essential polymers, in situ polymerization is limited because a suitable monomer–silicate solvent system is not always available, and it is not always compatible with current polymer-processing techniques [16, 17]. Baniasadi et al. [18] used this method to obtain exfoliated polypropylene-based nanocomposites using a bi-supported catalyst. Using cetyltrimethylammonium bromide as an organic modifier, Yuan et al. [19] used this method to prepare phenyl methyl silicone/organic MMT nanocomposites.

4.2.1.1.2 Solution blending

With the solution-blending technique, a solvent capable of swelling the clay and dissolving the polymer matrix is selected. A homogeneous three-component mixture of appropriate composition of the polymer, clay, and solvent is prepared through heating and mechanical and/or ultrasonic stirring. Depending on the interactions of the solvent and nanoparticles, the nanoparticle aggregates can be disintegrated in the solvent due to the weak van der Waals forces that stack the layers together [10]. Polymer chains can then be adsorbed onto the nanoparticles. The final step of the procedure involves removal of the solvent through evaporation or by precipitation in a nonsolvent. There is entropy gain associated with the removal of the solvent molecules, and it is thought to compensate the entropic loss of the intercalated polymer chains [20]. Therefore, it can be speculated that entropy-driven intercalation might be expected to occur even in the absence of an enthalpy gain due to favorable interactions between the macromolecules and the surface of the clay layers [21].

This technique is used especially with water-soluble polymers, such as polyvinyl alcohol (PVA) [22], poly-(acrylic acid) [23], poly(ethylene oxide) [24], poly(ethylene vinyl alcohol), to mention just a few. The polarity of such polymers is believed to contribute an enthalpy gain helping intercalation [25]. PVC has been used with organically modified Na-MMT to prepare nanocomposites using the method under discussion [10]. This procedure requires that the selected polymer be compatible with the selected solvent. There are examples of organic solvents and hydrophobic polymers such as polypropylene [26] that have also been given consideration. Notably, this method produces a high degree of intercalation only for certain polymer/clay/solvent systems, implying that for a given polymer, one has to find the right clay, organic modifier, and solvents.

4.2.1.1.3 Melt blending

The melt-blending procedure, also known as melt intercalation, came to prominence in the late 1990s [27]. This method involves the physical mixing of the polymer matrix and the clay in the molten state of the polymer. This process is attractive to researchers since this method is the most versatile and environmentally benign among all the methods of preparing PCNs. Nanocomposite synthesis via this method involves compounding and annealing (usually under shear) of a mixture of polymer and clay above the melting point of the polymer. During compounding and blending, the polymer melt diffuses into the cavities of the clay. This method, therefore, allows the processing of PCNs to be articulated directly from the precursors without using any solvent but ordinary compounding devices such as mixers and/or extruders.

Some polymers, for example ethylene vinyl acetate (EVA) [28] and thermoplastic polyurethane [29] and PCL [30], are thermoplastic polymers that have been used in the study of nanocomposite synthesis by melt intercalation. Wan et al. [31] investigated the effect of silicate modification and MMT content on the morphology development, relaxation behavior, and mechanical properties of the PVC/MMT nanocomposites. Similarly, Cabedo et al. [8] prepared EVOH-kaolinite nanocomposites by the melt intercalation.

Technically, the melt-blending method is much simpler and more straightforward compared to the methods discussed in the previous sections. In addition, even though relatively new, this method has more appealing advantages, which promise to greatly expand the commercial opportunities for PCN technology [16, 17]. One such advantage is that this approach does not use any organic solvent, and it is compatible with existing industrial polymer extrusion and blending processes [10]. The compatibility with existing thermoplastic polymer-processing techniques minimizes capital costs, and nonuse of organic solvents eliminates environmental concerns [16, 17]. In that respect, it is technically sound and suggests that the melt-blending method shifts PCN production costs downstream, thus giving end-use manufacturers some degree of freedom with regard to final product specifications (e.g., selection of polymer grade, choice of organoclay, level of reinforcement, etc.) [32]. At the same time, noninvolvement of solvents enhances the specificity for the intercalation of polymer in the clay galleries by eliminating the competing clay–solvent, polymer–solvent, and solvent–solvent interactions [32]. This should reduce hindrances in intercalation.

On the negative aspects, this method forms microcomposites or tactoids at higher clay loading as a result of clay agglomeration. Furthermore, this method employs thermoplastic polymers, which are generally hydrophobic, and this limits the application of the nanocomposites in water treatment. Finally, the compounding and extrusion processes are likely to amplify the range of particle shapes and sizes, principally when the clay is not uniformly exfoliated.

4.2.1.2 Polymer–clay nanocomposite structure

The improvement of the properties of the polymers is determined by the structure of the nanocomposite. The classically encountered structures are tactoid, intercalated, and exfoliated. The tactoid structure is mostly a consequence of intrinsic immiscibility between the polymer and the clay, leading to agglomeration of the clay in the polymeric matrix. The clay is not easily dispersed in most polymers due to their preferred face-to-face stacking in agglomerated tactoids [33]. Composites that consist of predominantly tactoid structures are also called microcomposites. Such composites can also be formed owing to the inability of the polymer to penetrate the galleries of the clay because the spacing is too narrow. In that case, the polymer envelopes the clay particles and so the dispersion of the clay particles in the polymer matrix is normally poor. The intercalated and exfoliated structures are illustrated in Fig. 4.3.

As the name suggests, intercalated structure is a consequence of intercalation. Here the polymer is located in the clay interlayers, expanding the clay structure, while still retaining some order between the platelets. Exfoliated structures, on the other hand, are a product of exfoliation whereby the original face-to-face structure of the clay platelets has been completely destroyed and single clay sheets are randomly dispersed in the polymer matrix. Exfoliated

PCNs are especially desirable because of their good macroscopic properties, mainly due to the large aspect ratio and interfacial area of the clay particles and optimal interaction between clay and polymer. However, it is very difficult to achieve this structure; as a result, a semi-exfoliated intercalated structure is formed. Importantly, the degree of exfoliation depends on many factors such as the chemical surface treatment of the clay particles, compatibility between layered silicates and polymer matrix, and the processing condition, which implies that the method of synthesis has an effect of the nanocomposite structure. The polarity of hydroxyl groups on the clay surface prevents nonpolar polymer chains from entering the galleries, hindering the formation of exfoliated nanocomposites. According to Vaia et al. [20], the interlayer structure of the organoclay should be optimized to maximize the configurational freedom of the functionalizing chains upon layer separation and to maximize potential interaction sites at the interlayer surface to achieve clay exfoliation.

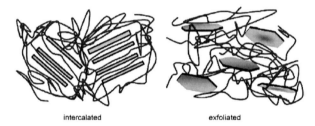

Figure 4.3   Schematic illustration of two different types of PCN structures sourced from Pavlidou et al. [32].

The nanocomposite structure is widely investigated using two complementary techniques, namely, X-ray diffraction spectroscopy (XRD) and transmission electron spectroscopy (TEM). However, of the two, XRD is the most frequently used to probe the nanocomposite structure due to its ease of use and accessibility. Typical XRD spectra indicating the presence of the discussed nanocomposite structures are shown in Fig. 4.4.

In this technique (XRD), the structure of the nanocomposite is identified by monitoring the position, shape, and intensity of the basal reflections from the distributed silicate layers on the polymer matrix. The structure of the silicates is not affected in microcomposites; therefore, the basal reflections do not change. In intercalated nanocomposite structures, there is an increase in the d-spacing of the clay, leading to a shift of the diffraction peak toward the lower angle, according to Bragg's law. The applicability of Bragg's law is another strong point of XRD, which allows the determination of the spaces between structural layers of the silicates as shown in the following formula:

$$\sin\theta = \frac{n\lambda}{2d}, \quad (4.1)$$

where λ is the wavelength of the X-ray radiation used in the diffraction experiment, d is the spacing between diffractional lattice planes, and θ is the measured diffraction.

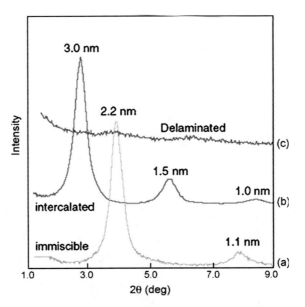

Figure 4.4 XRD spectra; adapted from Beyer [9].

This law is useful when characterizing intercalated nanocomposites due to the changes in the interlayer spacing to be determined. In contrast, the extensive layer separation associated with exfoliated structures interrupts the coherent layer stacking associated with the clay and causes a featureless diffraction pattern. As a consequence, exfoliated structures show no visible diffraction peaks in the XRD diffractograms with reference to the interlayer spacing of the clay. This could either be as a result of too large spacing between the layers or because the nanocomposite no longer presents any ordering. For example, it has been reported that interlayer spacing exceeding 8 nm, especially in the case of ordered exfoliated structure, does not exhibit visible diffraction peaks in the XRD diffractograms [32].

The disadvantage of the XRD technique is that it does not allude to the spatial distribution of the silicate layers or any structural inhomogeneities in nanocomposites. Additionally, this technique only uses basal reflection of the initial clay because some layered silicates initially do not exhibit well-defined basal reflections, so the peak-broadening and intensity decreases are difficult to study systematically [32]. Therefore, some researchers have used TEM, which allows a qualitative understanding of the internal structure and can directly provide information in real space, in a localized area, on morphology, and defect structures.

The TEM technique takes advantage of the fact that silicate layers consist of heavier elements (Al, Si, and O) than the interlayer and surrounding matrix (C, N, and H). The silicate layers, therefore, appear darker in bright-field images in microscopy [32]. A typical TEM micrograph depicting the different PCN structures is shown in Fig. 4.5.

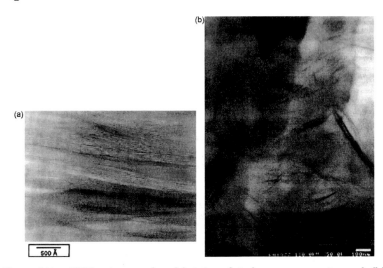

Figure 4.5 TEM micrographs: (a) intercalated nanocomposite and (b) exfoliated nanocomposite. Adapted from Alexandre et al. [34].

Figure 4.5a shows a picture of a PCN with intercalated structures as can be seen by the dark lines, and Fig. 4.5b shows exfoliated structures as seen by the dominant dark areas. The dark lines in Fig. 4.5 are cross sections of intercalated silicate layers [31]. The pictures further show that the silicates were well dispersed in the polymer matrix, which is an outstanding feature of TEM compared to XRD. TEM uses a very small sample compared to XRD, and so it is necessary to exercise special care to guarantee a representative cross section of the sample. Lastly, the samples have to undergo microtoming into thin sections for TEM analysis, and this may also result in an apparent redistribution of observed particle sizes even if all disk-like platelets were of the same size [31].

Additionally, differential scanning calorimetry (DSC) has the potential to be used to elucidate the PCN structure, particularly in intercalated nanocomposites. Free polymer chains have a certain degree of rotational and translational mobility, and this is normally determined by assessing the glass-transition temperature (Tg) of the particular polymer. The interactions between the intercalated chains of the polymer form and the layered silicate species greatly reduces the rotational and translational mobility of the polymer, which increases its Tg [32]. The augmentation of Tg is due to elevation of the energy threshold needed for the transition.

The structure of the PCN determines the morphology of the material. It is worth mentioning that the PCN morphology is widely studied using XRD, TEM, and DSC. In the context of XRD and DSC, PCN morphology is basically about investigating the changes in the crystallinity of the polymer matrix. It is essential to take into account any crystalline change in the polymer matrix in the presence of clays, especially for nanocomposites intended for applications in water treatment. It has been reported that crystalline polymers are considered to be impermeable to small molecules such as water [35]. Of the three techniques, XRD and DSC are the most useful when determining crystallinity, while TEM and scanning electron microscope (SEM) are ideal for interrogating the dispersion of clay particles. Zha et al. [13] used XRD and TEM to investigate the dispersion characteristics of polystyrene-block-polyisoprene-block-poly(2-vinylpyridine) (SI2VP triblock) copolymers prepared by solution blending and melt blending. XRD and TEM have also been used for assessing clay morphology and dispersion in polypropylene–clay nanocomposites prepared by in situ polymerization [18].

4.3 Polymer–Clay Nanocomposites in Heavy-Metal Removal from Water

4.3.1 Heavy-Metal Adsorption

Removal of pollutants such as heavy metals is a challenge especially in industrial effluents because of their threat to human and aquatic life [36]. PCNs have found use in water treatment where they are used in heavy-metal adsorption because of suitable clay properties such as large specific surface area, chemical and mechanical stability, layered structure, high cation-exchange capacity, to mention a few. In addition, clays can be regarded as possessing both Brönsted and Lewis types of acidity [2], and this acidity makes them suitable for the adsorption of heavy metals. The Brönsted acidity arises from two situations, first through the formation of H$^+$ ions on the surface, resulting from the dissociation of water molecules of hydrated exchangeable metal cations on the surface, as follows:

$$[M(H_2O)]_x^{n+} \rightarrow [M(OH)(H_2O)_{x-1}]^{(n-1)+} + H^+$$

Second, it can be as a result of a net negative charge on the surface due to the substitution of Si^{4+} by Al^{3+} in some of the tetrahedral positions and the resultant charge is counterbalanced by H$_3$O$^+$ cations. On the other hand, Lewis acidity can emanate from three scenarios: (1) through dehydroxylation of some Brönsted acid sites; (2) exposed Al^{3+} ions at the edges; and (3) Al^{3+} arising from the breaking of Si–O–Al bonds [37]. The resulting negative net charge is counterbalanced by exchangeable cations such as K$^+$, Ca^{2+}, and Mg^{2+}, adsorbed between the unit layers and around the edges, making it possible for clay to remove heavy metals from water through ion exchange, chemisorption, or physisorption.

Based on the aforementioned, it can be asserted that solution blending and in situ polymerization methods are not worth being considered in the preparation of PCNs aimed for water treatment because the products are in the form of beads. These beads have a lower surface area compared to the clay, and in this case the polymer hardly addresses the problem of complete recovery, but it can minimize sludge formation. Fan et al. [38] preferred coating chitosan with MMT for the removal of Cr^{6+}. Immobilizing the clay on the polymer by coating ensures that the polymer does not interfere with the properties that make clay particles ideal adsorbents for heavy metals. The melt-blending method, on the other hand, is capable of producing PCN strips. The advantages of the strips compared to beads and powder are discussed broadly in the following sections. Moreover, the solution and in situ polymerization methods are credited with producing intercalated structures. In the context of adsorption, the main purpose of a polymer in PCNs as applied in water treatment is to act as a support of the clay particles, which contain the adsorption sites. It is generally assumed that the polymer plays little or no role in the adsorption process. However, it is worth mentioning that depending on the type of polymer and the hydration state of the target heavy metal, the polymer can extract the pollutant metal from water through physical adsorption. The clay particles possess adsorption sites at the edges and surface [32], which imply that exfoliated structures are advantageous. Based on the force that is applied when preparing PCNs using the melt-blending procedure, more exfoliated structures can be obtained, as highlighted in a previous section. Very little has been reported on the removal of heavy metals from water using PCNs prepared by the melt-blending method, probably because hydrophobic polymers are involved.

The absorptive properties of nanocomposites can be influenced by the pH of the solution. For example, the adsorption of Pb^{2+} is reported to increase with an increase in pH from 3 to a peak at 4.8 for the nanocomposites prepared from EVA or PCL with bentonite clay [39]. The low Pb^{2+} uptake at low pH was reported to be due to competitive adsorption between Pb^{2+} and H_3O^+ as a result of high concentrations of H_3O^+. The adsorption equilibrium occurred at pH 4.8, an observation that was attributed to an increase in the concentration of $Pb(OH)^+$ as $OH-$ edges toward balancing H_3O^+. The hydrolyzed lead can undergo chemical adsorption with SiO_2 and Al_2O_3 in the clay. As can be seen in Fig. 4.1, the structure of 2:1 silicates such as bentonite as well as MMT consists of Al and Si atoms located on the edge sites and on the planar (internal) sites of the clay mineral. The atoms located on the edge sites are hydrolyzed to silanol (Si-OH) and aluminol (Al-OH). In MMT, adsorption can occur both at the edge sites, which leads to inner-sphere metal complexes, and at the planar (internal) sites of the clay mineral, which results in outer-sphere metal complexes. These unsaturated edge sites are much more reactive than the saturated basal sites [40]; the implication is that the hydrolyzed form of a heavy metal such as Pb^{2+} is less likely to be adsorbed. The negative charge should favor the removal of Pb^{2+}

in the hydrolyzed form, i.e., Pb(OH)$^+$. The adsorption efficiency of Pb(OH)$^+$ at higher pH is predominantly determined by the existing equilibrium between Pb(OH)$_2$ and Pb(OH)$^+$ PCN complex. However, at pHs higher than 10, the dominant species is Pb(OH)$_3^-$, which can only be repelled instead of being adsorbed on the negatively charged clay in the nanocomposite.

The adsorption of heavy-metal ions occurs in at least four stages that can have different kinetic rates [41]. First, the transport of the analyte involves the migration of Pb^{2+} from the bulk liquid to the composite particles. Since the nanocomposite particles would be enveloped by liquid film, the second stage of the adsorption process for heavy-metal ions involves penetration through the film to gain contact with the nanocomposite particles. In the next step, there will be intraparticle diffusion of the heavy-metal ions to reach the active sites of the composite. Finally, attachment will occur. It is imperative to identify the rate-determining step because it may control the kinetics of the adsorption process and therefore determine the removal rate of heavy-metal ions from the aqueous solution.

For adsorption experiments done under vigorous shaking, the migration of the heavy-metal ions from the liquid phase to the surface of the nanocomposite particles cannot be the rate-determining step. The interaction (binding) of heavy-metal ions with the nanocomposite may not be the rate-determining step because the adsorbate and adsorbent are hypothesized to bind instantly due to the opposite charges. Therefore, the rate-limiting step could either be the migration of heavy-metal ions through the nanocomposite pores (intraparticle diffusion) (in highly amorphous nanocomposite strips for PCNs prepared through melt-blending) or the liquid-film diffusion process. According to Alemayehua et al. [42], the following equation, known as the liquid-film model, can be used to predict if the liquid-film diffusion process is the rate-determining step in an adsorption process:

$$\ln(1-F) = -k_{fd}t, \qquad (4.2)$$

where F is a constant ($F = q_t/q_e$), k_{fd} is the adsorption rate constant, and t is time (h).

A linear plot of $-\ln(1-F)$ versus t passing through the origin indicates that the liquid-film diffusion is the rate-determining step of the adsorption process. According to Wang et al. [43] and Li et al. [44], the following equation can be applied to determine if intraparticle diffusion is the rate-limiting step in an adsorption process:

$$q_t = K_p t^{1/2} + C, \qquad (4.3)$$

where K_p is the intraparticle diffusion rate constant (in mg·g^{-1}h$^{1/2}$)

When the plot of q_t versus $t^{1/2}$ is linear and passes through the origin, the intraparticle diffusion process is the primary limiting mechanism [45].

4.3.1.1 Tailored morphology to enhance adsorption

As already mentioned in a previous section, the melt-blending method is advantageous over other methods of synthesis for the fabrication of adsorbents because it does not make use of any solvent, which minimizes hazardous environmental consequences and it is quick, automated, and easy to use while also compatible with conventional methods such as extrusion. This method is suitable for thermoplastics such as EVA, which is hydrophobic. The hydrophobic nature of such polymers can be advantageous with respect to stability under wet conditions. To ensure that more of the adsorbate readily reaches the filler (clay) for uptake, high filler content is required or alternatively, the composite strip should be made to be as thin as possible to expose more filler particles onto the surface. Applying the filler as a coating layer through dip coating might be ideal, but this would involve the use of chemicals, which is being avoided by opting for the melt-blending method when fabricating the adsorbents.

A safe method for modifying the morphology of PCN prepared by the melt-blending method for heavy metal removal is the use of water-soluble salt. Dlamini et al. [39] used two semicrystalline thermoplastic polymers EVA and PCL, and bentonite clay was selected for composite fabrication. The hydrophobic nature of EVA and PCL in water systems leads to stability of its chemical and mechanical properties, promoting its use in aqueous environments. The polymers were compatible with the melt-blending method, and the objective was to determine how the structural differences influence the adsorption of Pb^{2+} from water. Bentonite was selected because of its hydrophilic nature and net negative charge. Due to the hydrophobic nature of the composites, anhydrous Na$_2$SO$_4$ was used to create pores (free volume) in the nanocomposites. The pores were formed when the Na$_2$SO$_4$ was removed, indicating the absence of chemical interactions between the molten polymers and Na$_2$SO$_4$.

The free-volume pores were necessary to improve the contact ratio between bentonite particles and the Pb^{2+} ions to counter the adverse effects of composite hydrophobicity.

Figure 4.6 shows the SEM micrographs of the composites demonstrating that the use of Na$_2$SO$_4$ in the fabrication of bentonite/EVA and bentonite/PCL nanocomposites results in the formation of a number of pores on the surface of the nanocomposites.

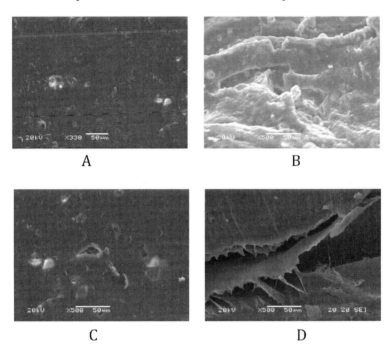

Figure 4.6 SEM images: (A) Na$_2$SO$_4$–EVA/bentonite surface morphology; (B) Na$_2$SO$_4$–EVA/bentonite fracture morphology; (C) Na$_2$SO$_4$–PCL/bentonite surface morphology; (D) Na$_2$SO$_4$–PCL/bentonite fracture morphology [39].

The pores were formed as a result of the removal of Na$_2$SO$_4$ indicating the absence of chemical interactions between the molten polymers and Na$_2$SO$_4$. On both nanocomposites, the pores appear as single isolated openings.

The Na$_2$SO$_4$–bentonite/EVA nanocomposite has a large average pore size compared to the Na$_2$SO$_4$–bentonite/PCL nanocomposite. For each nanocomposite, the pore size distribution was determined by analyzing the size of the pores shown in the SEM images. The

EVA nanocomposite has a noncircular and irregular pore structure compared to the PCL nanocomposite. It is possible that these pores might be disconnected. An inspection of the specimens on the opposite sides does not indicate pores with identical size and shape.

Figure 4.6 depicts the rough morphology of the Na_2SO_4-bentonite/EVA while Figure 6 shows that the Na_2SO_4-bentonite/EVA has a fractured morphology with a loosely held layered structure. A similar observation was made in our previous work on a comparative study of C20A/EVA and C20A/PCL nanocomposites. This was attributed to the penetration of PCL into the inter-gallery region of the clay. EVA is bulkier relative to PCL. Hence, steric hindrance is expected when EVA penetrates into the bentonite intergallery region. The fractured morphology is anticipated to complement the small pores in the PCL nanocomposite.

Preliminary results indicated that EVA/bentonite and PCL/bentonite nanocomposites can extract less than 10% of Pb^{2+} from a 200 mg/L Pb^{2+} solution. The low adsorption efficiency is attributed to low contact ratio between the adsorption active sites (in the clay) and the heavy metal. This is because the polymers are generally hydrophobic and the low clay loading of 3% (w/w). Therefore, Pb^{2+} adsorption was limited only to the clay particles that protrude on the surface of the nanocomposite. The low water sorption renders the clay particles embedded within the nanocomposite membrane useless. After adding and washing off Na_2SO_4, water sorption capacity increased as a result of the formation of large free-volume pores.

In another study, Dlamini et al. [46] blended EVA and polyurethane foam to improve the water sorption capacity and the porosity of PCN. Herein, polyurethane foam is being used to improve water penetration in EVA/bentonite composite fabricated via the melt-blending technique. EVA with 9% vinyl acetate is hydrophobic and therefore stable in a water environment. On the other hand, polyurethane has a high affinity for water.

Figure 4.7 shows typical SEM images of neat bentonite/EVA and bentonite/ polyurethane/EVA nanocomposites. The SEM image of bentonite/EVA nanocomposite depicts surface morphology ridges. No pores (both dead-end and continuous) are visible at the magnification used. After adding polyurethane, large voids were created, resulting from the expansion and opening of polyurethane during thermal blending. These could also be due to the immiscibility between the two polymers as polyurethane is hydrophilic while EVA is hydrophobic.

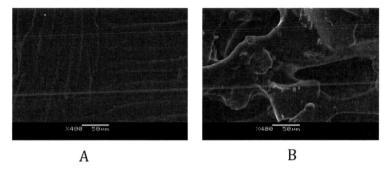

Figure 4.7 (A) Bentonite/EVA nanocomposites and (B) bentonite/polyurethane/EVA nanocomposites.

In water sorption tests, pure EVA and bentonite/EVA nanocomposite did not swell in water, while the polyurethane/EVA blend attained a maximum percentage of deionized water uptake of 2% mol. The nanocomposites were then tested for the removal of Pb^{2+} from water. The bentonite/polyurethane/EVA nanocomposite successfully removed 90% of Pb^{2+} from an aqueous solution with an initial concentration of 30 mg/L, while bentonite/EVA could only extract less than 10% of Pb^{2+} from water. This shows that polyurethane can improve the performance of the nanocomposite prepared by the melt-blending method in heavy-metal removal from water.

4.3.2 Heavy-Metal Retention by Granular Filtration

In granular filtration, water flows through granular material, while suspended or dissolved impurities in water are removed by attachment to the grains of a filter medium [47]. Filters are used for chemical and biological reactions in water treatment. Heavy metals can be removed from water using filters through chemical reaction by selecting an appropriate granular bed filter material. Granular bed filters are broadly used in potable water and wastewater treatments, with sand and activated carbon materials being the most common filter medium. An important property of a filter bed is that it should be easy to clean. After a while, filters should be periodically cleaned to remove suspended solids that fill the pores resulting in an increase in hydraulic pressure. Clay is a good adsorbent of heavy metals, but it is not a good filter bed material because of their very small sizes and flaky shapes, which result in very low bed porosity filter and limited grain resistance to attrition. This causes high hydraulic resistance, which is an unfavorable consequence, because it results in low water flux. The advantage of using filters to separate heavy metals from water is that there is no clogging of pores as compared to when they are used to remove organic and biological pollutants.

Polymer–clay nanocomposites in bead form with high adsorbent potential have a potential to be used in filtration operations. However, these are PCN prepared by the solution-blending or in situ polymerization methods because they are in bead form. Nanocomposites fabricated through the melt-blending method can be reconfigured to be suitable for use as granular filter material. For instance, Dlamini et al. [48] derived a heavy-metal adsorbent from the EVA/C20A nanocomposite through heat treatments of nonbiodegradable nanocomposites. The nanocomposite was burned under inert conditions and in air. The results are shown in Fig. 4.8.

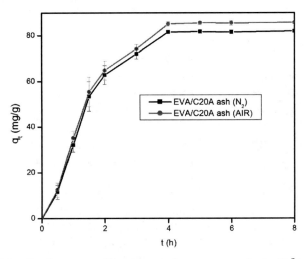

Figure 4.8 Performance of heat-treated nanocomposite in Pb^{2+} removal from water [48].

Apparently, there was little change in the adsorption properties of the adsorbent thermally treated in either air or N_2. A simple laboratory-scale setup can be designed to suite the application of

PCNs in column filtration as shown in Fig. 4.9. The two-layer filter bed is made of nanocomposite prepared by heat treatment in air atmosphere and PVA. The nanocomposite is responsible for removing the pollutant, while PVA is only there to increase the residence time of the water in the column. The PVA layer is more porous than the PCN layer because PVA has a larger particle size. This, therefore, means that the flow rate of water is higher in the PVA layer, so little attachment of pollutants can be expected.

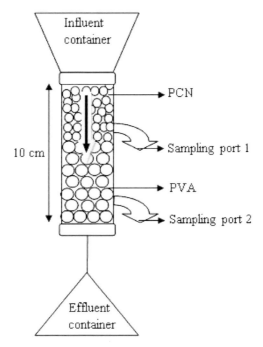

Figure 4.9 Possible setup for using PCNs in column filtration.

The transport of solutes in a filter bed is best discussed using the trajectory model approach. In this approach, there are three dominant processes governing the transport of particles: interception, sedimentation, and inertia. These processes are demonstrated in Fig. 4.10.

In interception, pollutants that do not reach the effluent container are intercepted by the filter bed particles. Interception occurs when a pollutant following a water streamline comes in contact with the filter bed particles. Sedimentation refers to the tendency for particles in suspension to settle out of the fluid streamline in which they are entrained and come to rest against a filter material. The pollutants settle out of the water streamline due to their motion in response to the forces acting on them. Heavy metals can deviate from the water trajectory and settle on the filter material as a result of attractive forces when the filter bed is made of negatively charged material such as nanocomposites made of clays. Interception applies to water pollutants larger than the open spaces between the filter particles. Particles smaller than the pores can be retained because of inertia. Inertia is the resistance of any physical object to a change in its state of motion or the tendency of an object to resist any change in its motion.

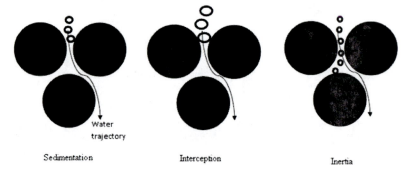

Figure 4.10 Transport of solutes in a filter bed.

4.3.3 Merits and Limitations of Polymeric Nanocomposites in Water Treatment

4.3.3.1 Merits

Nanocomposite brittleness is a drawback in some applications [34], but not in the removal of heavy metals from water. Nanocomposite brittleness allows water to permeate through the nanocomposite with ease ensuring that all adsorption active sites are accessible. The brittleness results from the disruption of the crystalline regions in the polymer. Nanocomposites prepared by the melt-blending method can be molded into strips of different sizes and shapes. This makes it relatively easy to recover the nanocomposites after heavy-metal capture in water. This is crucial because to obtain clean water the used adsorbent should be successfully removed from the water.

The size of the strip affords better handling and cleaning of the nanocomposite, compared to powder adsorbents such as activated carbon and clay. Additionally, embedding the adsorbents like clay in a polymer reduces the risk of sludge formation.

Other advantages of using clay materials as filler in nanocomposite earmarked for water treatment are low moisture resistance, which allows swelling of the fibers, and high heat resistance, which makes water treatment at higher temperatures possible. This allows the thermodynamic feasibility investigation of the adsorption of a heavy metal in a novel adsorbent. The thermodynamic parameters such as Gibbs free energy ($\Delta G°$), enthalpy ($\Delta H°$), and entropy ($\Delta S°$) can be determined to obtain a deeper insight into the adsorption of an adsorbate by the nanocomposite.

4.3.3.2 Limitations

Among the properties that make clay a suitable heavy-metal adsorbent is the specific high surface area. However, this property is suppressed by embedding the clay particles in a polymer despite the fact that surface area is an important parameter in adsorption technology. Furthermore, embedding the clay particles in a polymer can slow down the adsorption process because the adsorption active sites are concealed or enveloped by the polymer. This can be dominant in nanocomposites prepared by the melt-blending method, which uses hydrophobic polymers. Consequently, there are long equilibration periods in adsorption experiments using nanocomposites, compared to when the filler is used independently. One of the most reported drawbacks in composites is weight increase [34]. Weight increase tends to favor the settling of the nanocomposites in water, hence vigorous mechanical stirring would be required. Finally, depending on the polymer used in preparing the nanocomposite, erosion of support in desorption studies can be a serious problem. Desorption of the adsorbate from an adsorbent usually occurs under harsh conditions.

References

1. Trindade M.J., Dias M.I., Coroado J., Rocha F. (2009). Mineralogical transformations of calcareous rich clays with firing: A comparative study between calcite and dolomite rich clays from Algarve, Portugal, *Appl. Clay Sci.*, 42, 345–355.

2. Martin R.T., Bailey S.W., Eberl D.D., Fanning D.S., Guggenheim S., Kodama H., Pevear D.R., Srodon J., Wicks F.J. (1991). Report of the Clay Minerals Society Nomenclature Committee: Revised classification of clay materials, *Clays Clay Miner.*, 39, 333–335.

3. Maes N., Heylen I., Cool P., Vansant E.F. (1997). The relation between the synthesis of pillared clays and their resulting porosity, *Appl. Clay Sci.*, 12, 43–60.

4. Rousseaux D.D.J., Sclavons M., Godard P., Marchand-Brynaert J. (2010). Carboxylate clays: A model study for polypropylene/clay nanocomposites, *Polym. Degrad. Stab.*, 95, 1194–1204.

5. Zulfiqar S., Ahmad Z., Ishaq M., Sarwer M.I. (2009). Aromatic–aliphatic polyamide/montmorillonite clay composite materials: Synthesis, nanostructure and properties, *Mater. Sci. Eng. A*, 525, 30–36.

6. Mishra S.B., Luyt A.S. (2008). Effect of organic peroxides on the morphology, thermal and tensile properties of EVA/organoclay nanocomposites, *Express Polym. Lett.*, 2(4), 256–264.

7. Cho Y., Komarneni S. (2007). Synthesis of kaolinite from micas and K-depleted micas, *Clays Clay Miner.*, 55, 565–571.

8. Cabedo L., Giménez E., Lagaron J.M., Gavara R., Saur J.J. (2004). Development of EVOH-kaolinite nanocomposites, *Polymer*, 45, 5233–5238.

9. Beyer G. (2002). Nanocomposites: A new class of flame retardants for polymers, *Plast. Addit. Compd.*, 4(10), 22–27.

10. Madaleno L., Schjødt-Thomsen J., Pinto J.C. (2010). Morphology, thermal and mechanical properties of PVC/MMT nanocomposites prepared by solution blending and solution blending + melt compounding, *Comp. Sci. Technol.*, 70, 804–814.

11. Zhang D., Zhou C., Lin C., Tong D., Yu W. (2010). Synthesis of clay minerals, *Appl. Clay Sci.*, 50, 1–11.

12. Zanetti M., Lomakin S., Camino G. (2000). Polymer layered silicate nanocomposites. Macromol, *Macromol. Mater. Eng.*, 279, 1–9.

13. Zha W., Han C.D., Moon H.C., Han S.H., Lee D.H., Kim J.K. (2010). Exfoliation of organoclay nanocomposites based on polystyrene-block-polyisoprene-block-poly(2 vinylpyridine) copolymer: Solution blending versus melt blending, *Polymer*, 51, 936–952.

14. Lee J.L., Zeng C., Cia X., Han X., Shen J., Xu G. (2005). Polymer nanocomposite foams, *Comp. Sci. Technol.*, 65, 2344–2363.

15. Lingaraju D., Ramji K., Rao N.B.R.M., Lakshmi U.R. (2011). Characterization and prediction of some engineering properties

of polymer–Clay/Silica hybrid nanocomposites through ANN and regression models, *Procedia Eng.,* 10, 9–18.

16. Fornes T.D., Yoon P.J., Hunter D.L., Keskkula H., Paul D.R. (2002). Effect of organoclay structure on nylon 6 nanocomposite morphology and properties, *Polymer,* 43, 5915–5933.

17. Cho J.W., Paul D.R. (2001). Nylon 6 nanocomposites by melt compounding, *Polymer,* 42, 1083–1094.

18. Baniasadi H., Ramazani A.S.A., Nikkhah S.J. (2010). Investigation of in situ prepared polypropylene/clay nanocomposites properties and comparing to melt blending method, *J. Mater. Des.,* 31, 76–84.

19. Yuan X., Li X., Zhu E., Hu J., Cao S., Sheng W. (2010). Synthesis and properties of silicone/montmorillonite nanocomposites by in situ intercalative polymerization, *Carbohydr. Polym.,* 79, 373–379.

20. Vaia R.A., Giannelis E.P. (1997). Lattice model of polymer melt intercalation in organically modified layered silicates, *Macromolecules,* 30, 7990–7999.

21. Filippi S., Mameli E., Marazzato C., Magagnini P. (2007). Comparison of solution-blending and melt intercalation for the preparation of poly(ethylene-co-acrylic acid)/organoclay nanocomposites, *Eur. Polym. J.,* 43, 1645–1659.

22. Strawhecker K.E., Manias E. (2000). Structure and properties of poly(vinyl alcohol)/Na+ montmorillonite nanocomposites, *Chem. Mater.,* 12, 2943–2949.

23. Billingham J., Breen C., Yarwood J. (1997). Adsorption of polyamine, polyacrylic acid and polyethylene glycol on montmorillonite: An in situ study using ATR-FTIR, *Vibr. Spectrosc.,* 14, 19–34.

24. Malwitz M.M., Lin-Gibson S., Hobbie E.K., Butler P.D., Schmidt G. (2003). Orientation of platelets in multilayered nanocomposite polymer films, *J. Polym. Sci, Part B, Polym. Phys.,* 41, 3237–3248.

25. Qiu L., Chen W., Qu B. (2006). Morphology and thermal stabilization mechanism of LLDPE/MMT and LLDPE/LDH nanocomposites, *Polymer,* 47, 922–930.

26. Chiu F.-C., Chu P.-H. (2006). Characterization of solution-mixed polypropylene/clay nanocomposites without compatibilizers, *J. Polym., Res.,* 13, 73–78.

27. Shen Z., Simon G.P., Cheng Y.-B. (2002). Comparison of solution intercalation and melt intercalation of polymer–clay nanocomposites, *Polymer,* 43, 4251–4260.

28. Srivastava S.K., Pramanik M., Acharya H. (2006). Ethylene/vinyl acetate copolymer/clay nanocomposites, *J. Polym. Sci. Polym. Phys.*, 44, 471–480.

29. Finnigan B., Martin D., Halley P., Truss R., Campell K. (2004). Morphology and properties of thermoplastic polyurethane nanocomposites incorporating hydrophilic layered silicates, *Polymer,* 45, 2249–2260.

30. Di Y., Iannace S., Maio E.D., Nicolais L. (2003). Nanocomposites by melt intercalation based on polycaprolactone and organoclay, *J. Polym. Polym. Phys.*, 41, 670–678.

31. Wan C., Qiao X., Zhang Y., Zhang Y. (2003). Effect of different clay treatment on morphology and mechanical properties of PVC-clay nanocomposites, *Polym. Testing,* 22 453–461.

32. Pavlidou S., Papaspyrides C.D. (2008). A review on polymer–layered silicate nanocomposites, *Prog. Polym. Sci.,* 33, 1119–1198.

33. Ai Z., Cheng Y., Zhang L., Qiu J. (2008). Efficient removal of Cr (VI) from aqueous solution with Fe@Fe2O3 core-shell nanowires, *Environ. Sci. Technol.,* 42, 6955–6960.

34. Alexandre M., Dubois P. (2000). Polymer–layered silicate nanocomposites: Preparation, properties and uses of a new class of materials, *Mater. Sci. Eng. R,* 28, 1–63.

35. Picard E., Vermogen A., Gerard J.-F., Espuche E. (2007). Barrier properties of nylon 6-montmorillonite nanocomposite membranes prepared by melt blending: Influence of the clay content and dispersion state: Consequences on modeling, *J. Membr. Sci.,* 292, 133–144.

36. Farajzadeh M.A., Monji A.B. (2004). Adsorption characteristics of wheat bran toward heavy metal cations, *Sep. Purif. Technol.,* 38, 197–207.

37. Bhattacharyya K.G., Gupta S.S. (2008). Adsorption of a few heavy metals on natural and modified kaolinite and montmorillonite: A review, *Adv. Colloid Interface Sci.,* 140, 114–131.

38. Fan D.H., Zhu X.M., Xu M.R., Yan J.L. (2006). Adsorption properties of chromium (VI) by chitosan coated montmorillonite, *J. Bio. Sci.,* 6, 941–945.

39. Dlamini D.S., Mishra A.K., Mamba B.B. (2012). Adsorption Behaviour of Ethylene Vinyl Acetate and Polycaprolactone-Bentonite Composites for Pb^{2+} Uptake, *J. Inorg. Organomet. Polym.,* 22, 342–351.

40. Ahmad R.B., Mehrdad K., Mohammad H.N.F., Mohammad H.R., Mohammad Mohammad H.B. (2008). High temperature ablation of kaolinite layered silicate/phenolic resin/asbestos cloth nanocomposite, *J. Hazard. Mat.,* 150, 136–145.

41. Bystrzejewski M., Pyrzyńska K. (2011). Kinetics of copper ions sorption onto activated carbon, carbon nanotubes and carbon-encapsulated magnetic nanoparticles, *Coll. Surf. A: Physicochem. Eng. Aspects*, 377, 402–408.
42. Alemayehua E., Thiele-bruhn S., Lennartza B. (2011). Adsorption behaviour of Cr(VI) onto macro- and micro-vesicular volcanic rocks from water, *Sep. Purif. Technol.*, 78 (1), 55–61.
43. Wang X.S., Lu Z.P., Miao H.H., He W., Shen H.L. (2011). Kinetics of Pb (II) adsorption on black carbon derived from wheat residue, *Chem. Eng. J.*, 166, 986–993.
44. Li L., Liu F., Jing X., Ling P., Li A. (2011). Displacement mechanism of binary competitive adsorption for aqueous divalent metal ions onto a novel IDA-chelating resin: Isotherm and kinetic modelling, *Water Res.*, 45, 1177–1188.
45. Tofighy M.A., Mohammadi T. (2011). Adsorption of divalent heavy metal ions from water using carbon nanotube sheets, *J. Hazard. Mater.*, 185, 140–147.
46. Dlamini D.S., Mishra A.K., Mamba B.B. (2012). Morphological, transport and adsorption properties of ethylene vinyl acetate-polyurethane/bentonite clay composite, *J. Appl. Polym. Sci.*, 124, 4978–4985.
47. Ma´ rquez G.E., Ribeiro M.J.P., Ventura J.M., Labrincha J.A. (2004). Removal of nickel from aqueous solutions by clay-based beds, *Ceramics International*, 30, 111–119.
48. Dlamini D.S., Mishra A.K, Mamba B.B. (2012). Comparative study of EVA-Cloisite® 20A and heat-treated EVA-Cloisite® 20A on heavy-metal adsorption properties, *Water SA*, 38, 519–528.

Chapter 5

Role of Polymer Nanocomposites in Wastewater Treatment

Balbir Singh Kaith,[a] Saruchi,[a] Vaneet Thakur,[a] Ajay Kumar Mishra,[b] Shivani Bhardwaj Mishra,[b] and Hemant Mittal[b]

[a]*Department of Chemistry, National Institute of Technology, Jalandhar 144011, Punjab, India*
[b]*Department of Applied Chemistry, University of Johannesburg, Doornfontein 2028, South Africa*
bskaith@yahoo.co.in

Nanomaterials based on natural sources such as plants and organoclay are the new types of materials possessing diversified properties such as treatment of industrial effluent and other wastewater treatments. Polyaniline nanocomposites are used for the removal of sulfates. Perchlorates can be removed using grapheme-polypyrrole nanocomposites. Biodegradable nanocomposites have a great scope for various potential applications as high performance materials. At the end of life cycle, such materials are safely decomposed into carbon dioxide, water, and humus through microorganism activity. Nanocomposites not only act as effective sorbent for the removal of pollutants from wastewater but also act as an excellent support for

---

Nanocomposites in Wastewater Treatment
Edited by Ajay Kumar Mishra
Copyright © 2015 Pan Stanford Publishing Pte. Ltd.
ISBN 978-981-4463-54-6 (Hardcover), 978-981-4463-55-3 (eBook)
www.panstanford.com

the gold nanoparticles to remove Co^{2+} through catalytic oxidation process. Water treatment through nanocomposites has proven to be superior to any other water treatment technology.

5.1 Introduction

Most of the natural materials are present in the form of nano or micrometrically structured composites. The interaction between their structural elements depends on the internal structure and their physical and chemical properties. Nanocomposites are multiphase solid materials, possessing one of phases with dimension less than 100 nm. Such nanostructures can be synthesized by the nanoscale repeat between the different phases. Nanocomposites may include porous media, gels, colloids, and copolymers. Nanocomposites are the solid combination of bulk matrix and have nanodimension phase, which differ in properties because of the dissimilarities in structural chemistry. The mechanical, electrical, optical, thermal, electrochemical, and catalytic properties of nanocomposites always differ markedly from that of the originating materials. Since there is a presence of physical and chemical interactions between different components, composites establish complex physical and chemical equilibria masking their original properties. The reinforcing agents used in nanocomposites may be particles, sheets, or fibers, and there exists a high surface-to-volume ratio. Nanocomposite is a fast growing area of research. Significant efforts are made to control the nanoscale structures via different innovative synthetic approaches. The properties of nanocomposite materials depend not only on the properties of their individual components but also on their morphology and interfacial characteristics. Nanocomposites formed from metallic particles dispersed in polymer and ceramic or vitreous matrices have important application in the areas of catalysis and electronics. Nanocomposites have been observed to exhibit better and innovative characteristics in comparison to their micro counterparts [1–4].

Scarcity of drinking water is the worldwide problem especially in the developing countries with increasing population, and requirement for potable water is increasing with the same pace. Today the world is facing the formidable challenge of meeting out water requirement, which is fast depleting due to various reasons such as draught, population growth, and other compelling demands. Wastewater is widely recognized as one of the significant and reliable water sources. Due to industrialization, its production is increasing at an alarming rate. Therefore, reclamation of wastewater has the scope of balancing the water availability and water demand. It also assists in improving the ecosystem. Since clean water means water free from toxic chemicals and pathogens, wastewater reclamation technologies should be cheap, effective, and sustainable. Moreover, it should be efficient in removing pathogen from wastewater. In developing countries such as India, 80% of the diseases are due to contamination of drinking water. Nanocomposites play an important role in reclamation of wastewater [5–7].

Industrial activities such as paper manufacturing, mineral processing, petrochemical industries, and mining activities release a large amount of wastewater enriched with sulfates. Sulfates cause corrosion in sewer during wastewater discharge. Production of sulfide is a safety issue and also a maintenance problem as hydrogen sulfide is a poisonous gas, and when it is oxidized, it leads to the formation of SO_2 and ultimately gets converted into sulfuric acid causing corrosion of metals, concrete, and gas engine. Sulfide is a wastewater contaminant and also influences the amount of biochemical oxygen demand, chemical oxygen demand, and human health. It leads to serious problems during on-site anaerobic treatment of water. Different strategies have been suggested for the removal of toxicity from wastewater. Initially, sulfate reduction treatment is carried out followed by second-stage methanogensis. Different techniques such as sulfite precipitation using iron salt, pH control, and gas striping use of anaerobic and filtration in down-flow mode to remove sulfide from wastewater are used. One of the suitable methods for removing sulfide from wastewater is the surface adsorption process. Polyaniline nanocomposite is one of the suitable adsorbents for the removal of sulfide from wastewater [8–11].

Rocket fuels and fireworks release perchlorate (ClO_4^-) in the environment. Since perchlorate is highly soluble and nonreactive, it affects water quality and is found in underground water. Perchlorates have significant effects on human health. They can block the uptake of iodine in the thyroid gland and affects the production of thyroid hormone. So removal of perchlorate from water is essential. It is difficult to remove perchlorates from water by conventional separation technology. Electrically switched ion exchangers have been developed as a green separation technology for removing ClO_4^- ions from water [12–21]. Zhang et al. [3] synthesized a novel graphene polypyrole nanocomposite, which served as an excellent electrically switched ion exchanger for removal of perchlorates.

Manganese-oxide-supported phosphomolybdic acid nanocomposite was synthesized for degrading various types of carcinogenic dyes—for example, methylene blue, an organic dye hazardous for human beings and found commonly in industrial effluents. Fullerene-ZnO nanocomposites have been found to degrade organic dyes such as methyl orange, methylene blue, and rhodamine B [22–27]. This chapter is designed as a comprehensive source for polymer nanocomposites, their fundamental structures, manufacturing techniques, and applications especially in the wastewater treatment.

5.2 Types of Polymer Nanocomposites

There are three types of polymer nanocomposites: conventional nanocomposites, intercalated nanocomposites, and exfoliated nanocomposites.

5.2.1 Conventional Nanocomposites

Conventional nanocomposites are prepared by using polymer matrix with inorganic material reinforcement, but such composites have poor interaction, which leads to separation into discrete phases. Inorganic fillers are added in higher concentration to enhance the thermomechanical properties of polymer nanocomposites [28–30].

5.2.2 Intercalated Nanocomposites

In intercalated nanocomposites, bending and folding of platelets within matrix are observed. Intercalating of polymer chains into clay results in the intact positioning of clay particles; such microstructures are called intercalated chains, and nanocomposites formed in this way are called intercalated nanocomposites [34, 35].

5.2.3 Exfoliated Nanocomposites

In exfoliated nanocomposites, filler platelets get delaminated and converted into their primary nanostructure. The periodicity of platelet arrangement is totally lost, and platelets are far apart from each other. This occurs when the electrostatic interaction between the platelets is completely overcome by the polymer chain. Disappearance of the coherent X-ray diffraction from the distributed layer takes place [35, 36].

5.3 Methods of Preparation

Polymer nanocomposites can be synthesized in four different ways: melt compounding, in situ polymerization, bulk polymerization, and electrospinning.

5.3.1 Melt Compounding

Melt compounding is carried out at high temperature. High-molecular-weight polymers such as polystyrene and polyethylene are melted at high temperature followed by the addition of clay powder. To achieve uniformity, the filler is mixed thoroughly with the melt. Since this method requires high temperature condition, sometimes it results in modification and degradation of the samples. So these days this technique is not used for the preparation of polymer-based nanocomposites [37–40].

5.3.2 In situ Polymerization

The in situ polymerization method is used only when nanoparticles do not interfere with the polymerization reaction. Compatibility of the polymer matrix and filler is desired for the synthesis of polymer nanocomposites. Compatibility can be improved through suitable blending conditions such as temperature and their mixing intensity. Chemically modified fillers can be used to improve compatibility.

5.3.3 Bulk Polymerization

Bulk polymerization is used in case of liquid monomers. Bulk polymerization of the monomer is carried out in the presence of clay or any other filler. This leads to better interfacial contacts between the two phases and does not suffer from the problem of intercalation of high-molecular-weight polymer chains present inside the clay interlayers. Solvent can be used as a reaction medium to reduce the viscosity of the reaction medium and distribute the heat more uniformly. The monomer used and polymer formed should be soluble, and the filler used should be swellable in the chosen solvent. At the end of polymerization reaction, precipitation of the product takes place. Precipitates formed are collected and dried. Suspension and emulsion polymerization methods can also be used to generate polymer nanocomposites using styrene and methylmethacrylate as monomers.

5.3.4 Electrospinning

Electrospinning is a new technique for synthesizing polymer nanocomposites and is used for producing light-weight coatings, where nanoscale polymer fibers with high surface area are produced as substrates for polymer dispersion or polymer melt. Nanoparticles are incorporated into the polymer matrix, which increase the electrical conductivity of the material. The solvent gets evaporated from the polymer dispersion, while traveling through electrospinning substrate, and the polymer nanocomposites get deposited on the substrate [41–44].

5.4 Characterization

5.4.1 X-Ray Diffraction

X-ray diffraction is the most common technique used in the study of nanocomposites. The scattering of X-ray beam on nanocomposites may be in phase with one another, which leads to the formation of wavefront, resulting in constructive interference and consequent diffraction. Bragg's equation can be used to relate the wavelength of X-ray, glancing angle of incidence, and interplanar spacing of the crystal and state the condition necessary for diffraction. Ratanarat et al. synthesized poly(ethylene oxide) (PEO) montmorillonite (MMT) nanocomposites with various amounts of sodium montmorillonite

(Na-MMT) by the melt technique for the treatment of organic wastewater. Figure 5.1 exhibits the X-ray diffraction (XRD) of PEO, Na-MMT nanocomposites.

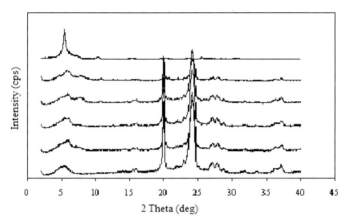

Figure 5.1 XRD spectra of PEO/Na-MMT nanocomposite with different amounts of Na-MMT [45].

5.4.2 Thermogravimetric Analysis

Thermogravimetric analysis (TGA) is widely used in the study of nanocomposites. Thermal analysis requires proper weighing of sample, loaded on a high precision thermobalance, and the sample undergoes a specific thermal program. Both test sample and reference pans are put under the same atmosphere. The purpose of the reference pan is to offset the weight of the test sample pan. Reference and sample pan are placed under high temperature condition, which undergo identical thermal cycles. The instrument detects the mass change under the influence of temperature, and the data is sent to the processor, which manipulates the data and plots in the form of thermogravimetric (TG) curves. Between the inflection points of the TG curve, the mass loss corresponding to the particular temperature range is measured. TG, differential thermal analysis, and derivative thermogravimetric analysis are very useful for the analysis of nanocomposites. The dehydroxylation of nanocomposites is another characteristic feature of the thermal behavior of nanocomposites. A lot of endothermic events occur within the temperature range of 100–500°C involving the loss of

–OH groups. Eksik et al. [46] synthesized oil-based polymer silver nanocomposites by photoinduced electron transfer and oil-based macromonomer from partial glycerides. The TGAs of these two are given in Fig. 5.2.

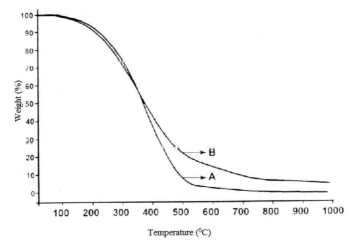

Figure 5.2   TGA of (a) oil-based macromonomer and (b) polymer silver nanocomposite [46].

5.4.3   Transmission Electron Microscopy

Transmission electron microscopy (TEM) is a high resolution technique that has the ability to evaluate the interior of the materials with appropriate resolution. TEM provides information concerning size and shapes of nanocomposites. This technique is based on the properties of electrons and is depicted by Eq. 5.1.

$$\lambda = \frac{h}{p} = \frac{h}{mv}, \qquad (5.1)$$

where $h = 6.626 \times 10^{-34}$ is Plank's constant and p, v, and m represent the momentum, speed, and mass of the electron, respectively. The high energy electrons can penetrate easily through the nanocomposite material, interacting inside the matrix and thereby projecting both the external and the internal surface of the material. Fornes et al. [47] synthesized organoclay nanocomposites of nylon-6 with three different molecular weight grades—high molecular weight (HMW), medium molecular weight (MMW), and low molecular weight

(LMW)—and was blended with 3 wt% MMT. The TEM images of HMW, MMW, and LMW are given in Fig. 5.3.

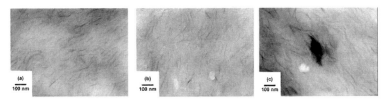

Figure 5.3   TEM images of melt-compounded nanocomposites containing, 3 wt% MMT based on (a) HMW, (b) MMW, and (c) LMW nylon-6 [47].

5.4.4   Scanning Electron Microscopy

The surface morphology of polymer nanocomposites is evaluated through the scanning electron microscopy (SEM) technique. SEM images have greater depth of field, which yield a characteristic 3D appearance. This 3D appearance is useful for understanding the surface morphology of the specimen sample. To record SEM images, a beam of electrons is required. These are accelerated toward the specimen sample using positive electrical potential. The electron beam is focused on the specimen sample using metal aperture and magnetic lenses. Interaction between the irradiated sample and electron beam is recorded. Kannan et al. [25] prepared manganese-oxide-supported polyoxometalate (POM) nanocomposite for the degradation of methylene blue dye. The SEM images of MnO_2 xerogel and manganese-oxide-based polyoxometalate nanocomposite is given in Fig. 5.4.

Figure 5.4   SEM images of (a) MnO_2 and (b) Mn_3O_4-POM [25].

5.5 Application of Polymer Nanocomposites

The following classes of nanocomposite materials are being used as functional materials for water purification: (a) dendrimers, (b) metal-containing nanoparticles, (c) zeolites, and (d) carbonaceous nanomaterials. These have a broad range of physiochemical properties that make them attractive as separation and reactive media for water purification. Characterization of the interactions of nanoparticles with bacterial contaminants by atomic force microscopy (AFM), TEM, and laser confocal microscopy shows considerable changes in the cell membranes, resulting in the death of the bacteria in most cases.

5.5.1 Dendrimers in Water Treatment

Reverse osmosis (RO) membranes have pore sizes of 0.1–1.0 nm and are very useful in retaining dissolved inorganic and organic solutes with molar mass below 1000 Da. Nanofilter (NF) membranes can be used for the removal of hardness, e.g., multivalent cations and organic solutes with molar mass between 1000 Da and 3000 Da and natural organic material. However, such operations required high pressures, whereas ultrafine (UF) membranes require low pressure (200–700 kPa). But the UF membranes are not very effective in removing dissolved organic and inorganic solute with molar mass below 3000 Da. Advancement in macromolecular science and technology such as the invention of dendritic polymers has resulted in the development of effective UF processes for the purification of water contaminated with toxic metal ions, radionuclide, organic and inorganic solutes, bacteria, and viruses. Dendrite polymers include random hyperbranched polymers, dendrigraft dendrons and dendrimers are relatively mono-dispersed and highly branched macromolecules with controlled composition and architecture. They contain three components: a core, interior branch cells, and terminal branch cell and tissue silver levels with 10% silver. Dendrimers mostly are symmetrical and spherical macromolecules. The interior may be similar or very different from the surface of the molecule [40, 49].

Their chemical or physical properties such as reactivity, complex or salt formation, and hydrophilicity or hydrophobicity can be varied. Diallo et al. [50] carried out dendron-enhanced ultrafiltration study.

They used polyaminoamine dendrimers with ethylene diamine as a core and terminal –NH$_2$ groups to recover Cu(II) ions from an aqueous medium. It has been observed that on mass basis, the Cu(II) binding capacities of polyaminoamine dendrimers were much larger and more sensitive to solution pH than those of linear polymers with amine groups [50].

5.5.2 Metal Nanocomposites

The most significant property of nanocomposites is that they have large surface area and can be further derivatized to increase their sorption capacity. They can be converted into metal ion or anion-selective sorbents. Interaction of nanocomposites with bacterial membrane, causing ultimate death of bacteria, which further can be proved through techniques such as AFM, TEM, and laser confocal microscopy. Photocatalytic nanocomposites can be used to destroy pesticides in the presence of UV radiation. Stoimenov et al. [51] observed that MgO nanoparticles and magnesium nanocomposites are effective biocides against Gram-positive and Gram-negative bacteria, e.g., *Escherichia coli* and *Bacillus megaterium* and bacterial spores. Magnesia nanoparticles, nanodots, or nanopowder possess high surface area and are typically 5–100 nm with specific surface area in the range of 25–50 m^{2}g^{-1}. The size ranges from 20 nm to 60 nm with 30–70 m^{2}g^{-1} surface area. Magnesium oxide nanoparticles have been found to absorb up to 20% of halogen molecules and have been observed to possess bactericidal property against different bacterial strains [51].

5.5.3 Zeolites

Zeolites are effective sorbents and ion exchange media for metal ions. Zeolites such as Na$_6$Al$_6$ Si$_{10}$O$_{32}$, 12H$_2$O have a high density of Na$^+$ ion exchange sites. They can be easily synthesized through hydrothermal activation of fly ash with low Si/Al ratio at 150 °C in 1.0–2.0 M NaOH solutions. Such zeolites can be used as ion exchange media for the removal of heavy metals from mine wastewaters and other industrial effluents. Alvarez-Ayuso et al. [53] reported the successful removal of Cr(III), Ni(II), Zn(II), Cu(II), and Cd(II) from metal electroplating wastewater. Nonporous ceramic oxides possessing large surface area (1000 m^{2}g^{-1}) and high density of sorption sites can be functionalized to increase the selectivity toward target pollutants. Zeolite nanoparticles can be prepared by laser-induced fragmentation techniques. Their formation takes place with the absorption of laser at impurities or defects within the zeolite microcrystals. This results in generating thermoelastic stress and mechanically fractures the microparticles into nanoparticle fragments. Large nanoparticles (>200 nm) are typically irregularly shaped crystals, whereas small nanoparticles (<50 nm) tend to be spherical and dense. Fragmentation and the amount of structural damage can be increased with increase in laser energy density, and the presence of strongly absorbing defects induce the plasma state with dramatical enhancement in temperature. Thus the optimal laser processing conditions are to be applied [52–54].

5.5.4 Carbonaceous Nanocomposites

Carbonaceous nanomaterials can serve as high capacity and selective sorbents for organic solutes in aqueous solutions. A number of polymers exhibiting antibacterial properties have been developed, e.g., soluble and insoluble pyridinium-type polymers involved in surface coating, azidated poly(vinyl chloride) used to prevent bacterial infection of medical devices, and polyethylene glycol polymers to prevent initial adhesion bacteria to the biomaterial surfaces. Bactericidal activity of polycationic agents is related to the absorption of positively charged nanostructures onto negatively charged cell surfaces of the bacteria. This process is responsible for the increase of cell permeability and destruction of bacteria cell membranes. Cross-linked polycations are prepared as nanoparticles. These are formed from polyethylene imine (PEI) by cross-linking and alkylation followed by methylation to increase the degree of amino group substitution. Because of the presence of positive charge and hydrophobicity, PEI nanoparticles have attracted attention as possible antimicrobial agents. PEI nanostructures can be evaluated for antibacterial action as a function of hydrophobicity, molecular weight, particle size, and charge present [55–59].

5.6 Conclusion

Nanocomposites based on plant and natural materials are new types of materials possessing diversified properties. Biodegradable nanocomposites have a lot of future scope for various potential applications such as water treatment and removal of toxic dyes from textile and leather industrial effluents. The composites based on natural materials get easily decomposed into carbon dioxide, water, and humus through microbial activity without any harm to the environment. Such high performance nanocomposites have great potential in industrial applications as well as potable water treatment, e.g., use of polyaniline and graphene polypyrrole nanocomposites for the removal of sulfate and perchlorate ions from wastewater, respectively. These nanocomposites are not only an effective sorbent for the removal of pollutants from wastewater but also an excellent support for gold nanoparticles to remove Co^{2+} through catalytic oxidation process. Water treatment through nanocomposites has proven to be superior to any other method.

References

1. Dhermendra, K., Tiwari, J., Behari, and Prasenjit, S. (2008). Application of nanoparticles in wastewater treatment, *World Appl. Sci. J.,* **3** (3), pp. 417–433.
2. US Bureau of Reclamation and Sandia National Laboratories. (2003). Desalination and water purification technology roadmap a report of the executive committee Water Purification.
3. Zhang, W., Zou, L., and Wang, L. (2009). Photocatalytic TiO_2/adsorbent nanocomposites prepared via wet chemical impregnation for wastewater treatment: A review, *Appl. Catal. A-Gen.,* **371**, pp. 1–9.
4. http://en.wikipedia.org/wiki/Nanocomposite.
5. World Health Organization. (1996). Guidelines for drinking water quality. Geneva: WHO, **2**.
6. US Environmental Protection Agency. (1998). Microbial and disinfection by-product rules. Federal Register, **63**, pp. 69389–69476.
7. US Environmental Protection Agency. (1999). Alternative disinfectants and oxidants guidance manual. EPA Office of Water Report, 815-R-99-014.

8. Mazyar, S.B., Vahid, B., and Faeqeh, P. (2011). Preparation of polyaniline nanocomposite for removal of sulfate from wastewater, in *2nd International Conference on Chemistry and Chemical Engineering IPCBEE*, IACSIT Press, Singapore, **14**.

9. Wiessner, A., Kappelmeyer, U., Kuschk, P., and Kastner, M. (2005). Sulfate reduction and the removal of carbon and ammonia in a laboratory-scale constructed wetland, *Water Res.*, **39**, pp. 4643–4650.

10. Tait, S., Clarke, W.P., Keller, J., and Batstone, D.J. (2009). Removal of sulfate from high-strength wastewater by crystallization, *Water Res.*, **43**, pp. 762–772.

11. Marina, N.S.A., Giovanni, P., Maria, L.F., Maria, C.G., Sedat, Y., Vincenzo, A., Mario, P., and Francisco, M. (2009). Self-assembled titania–silica–sepiolite based nanocomposites for water decontamination, *J. Mat. Chem.*, **19**, pp. 2070–2075.

12. Wu, J., Liu, C., Chu, K.H., and Suen, S. (2008). Removal of cationic dye methyl violet 2B from water by cation exchange membrane, *J. Membrane Sci.*, **309**, pp. 239–245.

13. Gupta, V.K., Ali, I., and Saini, V.K. (2004). Removal of rhodamine B, fast green and methylene blue from wastewater using red mud, an aluminum industry waste, *Ind. Eng. Chem. Res.*, **43**, pp. 1740–1747.

14. Kansal, S.K., Singh, M., and Sud, D. (2007). Studies on photodegradation of two commercial dyes in aqueous phase using different photocatalysts, *J. Hazard. Mater.*, **141**, pp. 581–590.

15. Kuo, W.S., and Ho, P.H. (2001). Solar photocatalytic decolorization of methylene blue in water, *Chemosphere*, **45**, 77–83.

16. Ilisz, I., Dombi, A., Mogyorosi, K., Farkas, A., and Dekany, I. (2002). Removal of 2-chlorphenol from water by adsorption combined with TiO_2 photocatalysis, *Appl. Catal. B. Environ.*, **39**, pp. 247–256.

17. http://www.eolss.net/Sample-Chapters/C07/E6-144-32-00.pdf.

18. Sheng, Z., Yuyan, S., Jun, L., Ilhan, A.A., and Yuehe, L. (2011). Graphene polypyrrole nanocomposite as a highly efficient and low cost electrically switched ion exchanger for removing ClO_4^- from wastewater, *Appl. Mat. Interf.*, **3**, pp. 3633–3637.

19. Urbansky, E.T., Magnuson, M.L., Kelty, C.A., and Gu, B. (2000). Brown GM: Comment on 'Perchlorate identification in fertilizers' and the subsequent addition/correction, *Environ. Sci. Technol.*, **34**, pp. 4452–4453.

20. Lin, Y.H., Cui, X.L., and Bontha, J. (2006). Electrically controlled anion exchange based on polypyrrole and carbon nanotubes nanocomposite for perchlorate removal, *Environ. Sci. Technol.*, **40**, pp. 4004–4009.

21. Crump, K.S., and Gibbs, J.P. (2005). Benchmark calculations for perchlorate from three human cohorts, *Environ. Health Perspect.*, **113**, pp. 1001–1008.
22. Sung, K.H., Jeong, H.L., Bum, H.C., and Weon, B.K. (2011). Preparation of a [70]fullerene-ZnO nanocomposite in an electric furnace and photocatalytic degradation of organic dyes. *J. Ceram. Process. Res.*, **12**, pp. 212–217.
23. Hasobe, T., Hattori, S., Kamat, P.V., and Fukuzumi, S. (2006). Supramolecular nanostructured assemblies of different types of porphyrins with fullerene using TiO_2 nanoparticles for light energy conversion. *Tetrahedron*, **62**, pp. 1937–1946.
24. Zhu, S., Xu, T., Fu, H., Zho, J., and Zhu, Y. (2007). Synergetic effect of Bi_2WO_6 photocatalyst with C_{60} and enhanced photoactivity under visible irradiation, *Environ. Sci. Technol.*, **41**, pp. 6234–6239.
25. Kannan, R., Gouse, P.S., Obadiah, A., and Vasanthkumar, S. (2011). MNO_2 supported pom–a novel nanocomposite for dye degradation, *Dig. J. Nanomater. Bios.*, **6**, pp. 829–835.
26. Moore, A.T., Vira, A., and Fogel, S. (1999). Biodegradation of trans-1,2-dichloroethylene by methane-utilizing bacteria in an aquifer simulator, *Environ. Sci. Technol.*, **23**, pp. 403–406.
27. Albanis, T.A., Hela, D.G., Hela, T.M., and Danis, T.M. (2000). Removal of dyes from aqueous solutions by adsorption on mixtures of fly ash and soil in batch and column techniques. *Global Nest: Int. J.*, **2**, pp. 237.
28. Lagaly, G. (1986). Interaction of alkyl amines with different types of layered compounds, *Solid State Ionic*, **22**, pp. 43–51.
29. Lagaly, G., and Beneke, K. (1999). Intercalation and exchange reactions of clay minerals and non-clay layer compounds, *Colloid Polym. Sci.*, **269**, pp. 1198–1211.
30. Osman, M.A., Seyfang, G., and Suter, U.W. (2000). Two-dimensional melting of alkane monolayers ionically bonded to mica, *J. Phys. Chem. B.*, **104**, pp. 4433–4439.
31. Osman, M.A., Ernst, M., Meier, B.H., and Suter, U.W. (2002). Structure and molecular dynamics of alkane monolayers self-assembled on mica platelets, *J. Phys. Chem. B.*, **106**, pp. 653–662.
32. Mittal, V. (2008). Epoxy-vermiculite nanocomposite as gas permeation barrier, *J. Comp. Mater*, **42**, pp. 2829–2839.
33. Mittal, V. (2008). Mechanical and gas permeation properties of compatibilised polypropylene-layered silicate nanocomposites, *J. Appl. Polym. Sci.*, **107**, pp. 1350–1361.

34. Kornmann, X., Linderberg, H., and Bergund, L.A. (2001). Synthesis of epoxy–clay nanocomposites: Influence of the nature of the curing agent on structure, *Polymer*, **42**, pp. 4493–4499.
35. Hasmukh, A.P., Rajesh, S.S., Hari, C.B., and Raksh V.J. (2006). Nanoclays for polymer nanocomposites, paints, inks, greases and cosmetics formulations, drug delivery vehicle and wastewater treatment, *Bull. Mater. Sci.*, **29**, pp. 133–145.
36. Ray, S.S., and Okamoto, M. (2003). Polymer/layered silicate nanocomposite: A review from preparation to processing, *Prog. Polym. Sci.*, **28**, pp. 1539–1641.
37. Dennis, H.R., Hunter, D., Chang, D., Kim, S., and Paul, D.R. (2001). Effect of melt processing condition on the extent of exfoliation in organoclay-based nanocomposites, *Polymer*, **42**, pp. 9513–9522.
38. Vaia, R.A., Jant, K.D., Kramer, E.J., and Giannelis, E.P. (1996). Microstructural evaluation of melt-intercalated polymer-organically modified layered silicate nanocomposites. *Chem. Mater.*, **8**, pp. 2628–2635.
39. Rehab, A., and Salahuddin, N. (2005). Nanocomposite materials based on polyurethane intercalated into montmorillonite clay, *Mater. Sci. Eng. A.*, **399**, pp. 368–376.
40. Burnside, S.D., and Giannelis, E.P. (1995). Synthesis and properties of new poly(dimethylsiloxane) nanocomposites, *Chem. Mater.*, **7**, 1597–1600.
41. Usuki, A., Kawasumi, M., Kojima, Y., Okada, A., Kurauchi, T., and Kamigaito, O.J. (1993). Swelling behavior of montmorillonite cation exchanged for v-amino acids by e-caprolactam, *Mater. Res.*, **8**, pp. 1174.
42. Alexandre, M., and Dubois, P. (2000). Polymer-layered silicate nanocomposites: Preparation, properties and uses of a new class of materials, *Mater. Sci. Eng. Rep.*, **28**, pp. 1–63.
43. Rehab, A., and Salahuddin, N. (2005). Nanocomposite materials based on polyurethane intercalated into montmorillonite clay, *Mater. Sci. Eng. A*, **399**, pp. 368–376.
44. Hussain, F., Dean, D., and Haque, A. 2002. Structures and characterization of organoclay-epoxy-vinyl ester nanocomposite, in *ASME 2002 International Mechanical Engineering Congress and Exposition*, doi:10.1115/IMECE2002-33552.
45. Ratanarat, K., Nithitanakul, M., Martin, D.C., and Magaraphan, R. (2003). Polymer-layer silicate nanocomposites: Linear peo and highly branched dendrimer for organic wastewater treatment, *Rev. Adv. Mater. Sci.*, **5**, pp. 187–192.

46. Eksik, O., Tasdelen, M.A., Errciyes, A.T., and Yagci, Y. (2009). In situ synthesis of oil-based polymer/silver nanocomposite by photoinduced electron transfer and free radical polymerization processes, *Compos. Interface.*, **17**, pp. 357–369.
47. Fornes, T.D., Yoon, P.J., Keskkula, H., and Paul, D.R. (2001). Nylon 6 nanocomposites: The effect of matrix molecular weight, *Polymer*, **42**, pp. 9929–9940.
48. Zeman, L.J., and Zydney, A.L. (1996). *Microfiltration and Ultrafiltration: Principles and Applications*, Marcel Dekker, New York.
49. Frechet, J.M.J., and Tomalia, D.A. (2001). *Dendrimers and Other Dendritic Polymers*, Wiley and Sons, New York.
50. Diallo, M.S., Christie, S., Swaminathan, P., Johnson, J.H., and Goddard, W.A. (2005). Dendrimer enhanced ultra-filtration recovery of Cu(II) from aqueous solutions using Gx-NH2-PAMAM dendrimers with ethylene diamine core, *Environ. Sci. Technol.*, **39**, pp. 1366–1377.
51. Stoimenov, P.K., Klinger, R.L., Marchin, G.L., and Klabunde, K.J. (2002). Metal oxide nanoparticles as bactericidal agents, *Langmuir*, **18**, pp. 6679–6686.
52. Brittany, L., Carino, V., Kuo, J., Leong, L., and Ganesh, R. (2006). Adsorption of organic compounds to metal oxide nanoparticles (Conference presentation is part of: *General Environmental*).
53. Alvarez, A.E., Sanchez, A.G., and Querol, X. (2003). Purification of metal electroplating wastewaters using zeolites, *Water Res.*, **37**, pp. 4855–4862.
54. Nichols, W.T., Kodaira, T., Sasaki, Y., Shimizu, Y., Sasaki, T., and Koshizaki, N. (2006). Zeolite LTA nanoparticles prepared by laser-induced fracture of zeolite microcrystals, *J. Phys. Chem.*, **110**, pp. 83–89.
55. Li, G. (2000). A study of pyridinium-type functional polymers. Behavioral features of the antibacterial activity of insoluble pyridinium-type polymers, *J. App. Pol. Sci.*, **78**, pp. 676–684.
56. Lakshmi, S., Kumar, S.S.P., and Jayakrishnan, A. (2002). Bacterial adhesion onto azidated poly (vinyl chloride) surfaces. *J. Biome. Mat. Res.*, **61**, pp. 26–32.
57. Lin, J. (2002). Bactericidal properties of flat surfaces and nanoparticles derivatized with alkylated polyethylene imines, *Biotec. Prog.*, **18**, pp. 1082–1086.
58. Park, K.D. (1998). Bacterial adhesion on PEG modified polyurethane surfaces, *Biomaterials*, 1998, **19**, 851–859.
59. Graveland, B.J.F., and Kruif, C.G. (2006). Unique milk protein based nanotubes: Food and nanotechnology meet, *Trends Food Sci. Technol.*, **17**, pp. 196–203.

Chapter 6

Nanoparticles for Water Purification

Pankaj Attri,[a,b] Rohit Bhatia,[b] Bharti Arora,[c] Jitender Gaur,[d] Ruchita Pal,[e] Arun Lal,[b] Ankit Attri,[d] and Eun Ha Choi[a]

[a]*Plasma BioScience Research Center/Department of Electrophysics, Kwangwoon University, 20 Kwangwoon-ro, Nowon-gu, Seoul 139-701, Korea*
[b]*Department of Chemistry, University of Delhi, Delhi 110007, India*
[c]*Department of Applied Science, ITM University, Sector 23(A), Gurgaon, Haryana 122017, India*
[d]*J & S Research and Innovations, New Delhi 110092, India*
[e]*Advanced Instrumentation Research Facility, Jawaharlal Nehru University, New Delhi 110067, India*
chem.pankaj@gmail.com

According to the United Nations, the world demand for clean water in the last century has increased sevenfold due to the quadruple increase in world population [1]. This has resulted in the lack of clean water availability to nearly 35% of world population and the majority of them reside in countries with low development index. Due to the nonavailability of clean water, combined with low sanitation level in these parts of the world, there have been widespread waterborne diseases, and it has been estimated that every year nearly 3.4 million deaths occur. In addition, the majority of these cases include children under the age of 5 years [2]. Because of the important role of clean water in human development, the United Nations has declared the decade 2005–15 as the decade for "water for life" [3].

Nanocomposites in Wastewater Treatment
Edited by Ajay Kumar Mishra
Copyright © 2015 Pan Stanford Publishing Pte. Ltd.
ISBN 978-981-4463-54-6 (Hardcover), 978-981-4463-55-3 (eBook)
www.panstanford.com

The conventional water purification technology involves coagulation, flocculation, sedimentation, filtration, chlorination, and ozonation as an essential step to purify water from microbial and chemical contaminations [2]. Coagulation and flocculation steps remove suspended particles such as clay, silt, bacteria, and organic matter from the contaminated water, but it also leads to the removal of beneficial minerals from the water, thereby decreasing the overall quality of water. The sedimentation and filtration steps help in the removal of floc and sludge generated in the first step, but one needs to check regularly condition of sedimentation bed; the algal growth and filters are needed to be replaced at regular intervals to have an effective operational level, thereby increasing the overall cost of water purification [3]. Chlorination and ozonation help in the removal of microbes present in water, which are mainly responsible for waterborne diseases [4]. As chlorine and ozone are well-known strong oxidizing agents, they react with natural organic matter present in water to generate disinfectant by-products (DBPs) such as trihalomethanes, haloacetic acids, haloacetonitriles, bromate, and chlorite [5, 6]; many of these by-products are classified as potential carcinogens by the World Health Organization [7]. Due to these drawbacks in conventional water purification, there is an urgent need to call for a technology that can overcome these drawbacks and allow easy access to potable clean water.

In view of this, nanotechnology has emerged as an alternative to the conventional water purification technology for effective removal of chemical and microbial contaminations due to their large surface-area-to-volume ratio, enhanced catalytic properties, high conductivity, antimicrobial properties, and self-assembly on surfaces [8–10]. Currently, nanotechnology-enabled water purification technology mainly focuses on three major areas for purifying water to improve its quality:

1. Removal of pollutants by absorption
2. Catalytic degradation of organic pollutants
3. Disinfection of microbial contaminations

For better performance of nanotechnology for water purification, metal and metal oxide nanomaterials such as silver (Ag), zero-valent iron, Fe_3O_4, and TiO_2 nanoparticles are being used for achieving the above-mentioned goals.

Among all the nanoparticles, silver (Ag) nanoparticles are most studied nanoparticles in the history nanotechnology. Since the ancient times, Ag metal has been known for its medicinal properties especially the antibacterial properties for the effective treatment of bacterial infections. People from the Phoenicians civilization used Ag-coated bottles to store water and other liquids to protect them from microbes [11]. It has been reported that the antimicrobial properties of Ag have been enhanced at the nanolevel due to their high surface-area-to-volume ratio and the unique chemical and physical properties [12, 13]. Ag nanoparticles are generally smaller than 100 nm and have around 15,000–20,000 Ag atoms [14]. Besides being bactericidal against the broad spectrum of bacteria, including the multidrug resistant strains, Ag nanoparticles have shown to be effective against common pathogenic fungi Aspergillus, Candida, and Saccharomyces [15]. Further, Ag nanoparticles in the range of 5–20 nm have shown to be effectively inhibiting the replication of HIV-1 virus [16, 17]. Due to these broad ranges of antimicrobial properties combined with its ability to produce no DBPs, Ag nanoparticles have been effectively used as a disinfectant for the purification of water.

The Ag nanoparticles have been synthesized from different methods such as chemical, physical, and biological methods. Chemical reduction has been the most common method used for the synthesis of Ag nanoparticles as a stable colloidal suspension in aqueous or organic solvents [18, 19]. Commonly used reducing agents used for synthesis of Ag nanoparticles include sodium borohydride, sodium citrate, sodium ascorbate, and elemental hydrogen [20–22]. During the synthesis of Ag nanoparticle, Ag ion is reduced to Ag atom (Ag^0), which then agglomerates into oligomeric clusters and these clusters then combine to form the Ag nanoparticles [23]. Ag nanoparticle generally appears as yellow in color (Fig. 6.1) and absorbs intensely in the range 300–400 nm. The size of nanoparticle depends on the strength of the reducing agent. Reducing agents such as sodium borohydride lead to the formation of particle with smaller size, which are monodispersed in nature, whereas weaker reducing agents have slow rate of reduction thereby resulting in the formation of particle with larger size [24].

Silver nanoparticles have also been synthesized by physical methods such as evaporation-condensation and laser ablation. In the laser ablation method, bulk material in solution has been used to synthesize nanoparticles. The size and morphology of nanoparticle synthesis using laser ablation depend on factors such as the wavelength of the laser used, the duration of the laser pulses, the duration of ablation, and the liquid medium [26–29]. The major advantages of using the physical method over the chemical method for nanoparticle synthesis have been that they are free from any chemical contaminations and one can have better control over the size of nanoparticle.

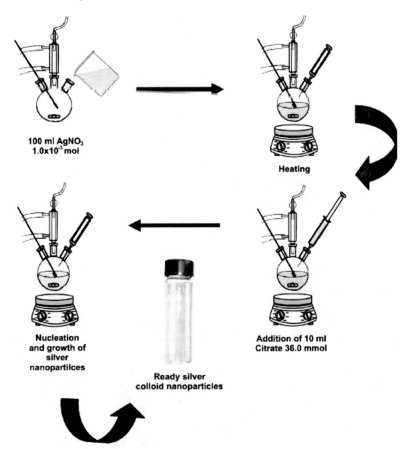

Figure 6.1 Synthesis of Ag nanoparticles using sodium citrate [25].

Besides chemical and physical approach, a number of authors have also used biological approach to carry out the synthesis of Ag nanoparticles. This has been a greener approach to synthesize

Ag nanoparticles without using any harsh, toxic, and expensive substances. In this approach, metal ion has been produced with the help of naturally occurring reducing agents present such as enzymes/proteins, amino acids, polysaccharides, etc. [30, 31]. Ag nanoparticles have also been synthesized using the supernatant extract from the culture medium of bacteria such as *Bacillus licheniformis, Bacillus subtilis, Klebsiella pneumonia, Escherichia coli,* and *Enterobacter cloacae* [32–34]. Similarly, extracellular extracts from fungi such as *Fusarium oxysporum, Fusarium acuminatum, Phanerochaete chrysosporium, Aspergillus fumigatus, Cladosporium cladosporioides, Penicillium fellutanum,* and *Coriolus versicolor* have been used to synthesize Ag nanoparticles [36–38]. Besides extracellular extract from fungi and bacteria, a number of authors have reported the synthesis of Ag nanoparticles of enhanced stability and microcidal actions using plant extracts such as *Camellia sinensis, Cymbopogon flexuosus, Datura metel,* and *Nelumbo nucifera* [39–41].

The methods discussed above for the synthesis of Ag nanoparticles can further be used for water purification. Ag nanoparticles impregnated or surface-coated membranes have been widely used in commercial water purification system (Aqua Pure, Kinetico, and QSI-Nano) to remove the 99.9% pathogen present in water [42]. Besides removal of pathogens from water, it has been shown that Ag nanoparticles can effectively remove halogenated pollutant present in water [43, 44]. It has been shown that Ag nanoparticles react with halocarbons resulting in the formation of metal halide on further reductive dehalogenation. There has been an advantage of this reaction: Only amorphous carbon is generated as by-product and the reaction occurs to completion at room temperature. Also, Ag nanoparticles have been effectively used to remove pesticides present in water by degrading.

Another application of Ag nanoparticles for purifying polluted water has been the removal of heavy metals present in water. It has been shown that Ag nanoparticles can chemisorb metal cations and lead to sequestration of heavy metals. A blue shift has been observed in surface plasmon coupled with a decrease in intensity when Ag nanoparticles were allowed to interact with Hg^{2+} cation. This has indicated that partial oxidation of Ag took place due to incorporation of mercury to Ag nanoparticles [45]. Similarly, it has been shown that Cd^{2+} can interact with Ag nanoparticles and can also be removed from water [46]. Recently, nanocomposites of Ag nanoparticle with chitosan and alginate have been used for the disinfection of water. Nanocomposites of Ag nanoparticles with alginate have been synthesized using three different methods; these beads were used as a filler material for the simultaneous filtration-disinfection of water [47]. Ag nanoparticle nanocomposites with chitosan have been fabricated with the help of microwave irradiation, and it has been used to remove the pesticides from water for endpoint use [48]. These above-mentioned applications of Ag nanoparticle help in water purification.

Other important nanoparticles used for the purification of water include nanoscale zero-valent iron (nZVI). In recent years, nZVI particles have emerged as one of the promising candidates for groundwater purification [49]. Zero-valent iron (Fe^0) has been recognized as an excellent electron donor, regardless of its particle size. Fe in its zero-valent state exhibits a strong tendency to release electrons in aquatic environments

$$Fe^0 \rightarrow Fe^{2+} + 2e^-$$

Zero-valent iron can also react to form the redox couples with other environmentally significant electron acceptors such as hydrogen ions, dissolved oxygen (DO), nitrate, and sulfate [50].

The first generation nanoscale zero-valent iron nanoparticles have been synthesized by reducing the iron chloride with sodium borohydride; reaction is shown in the following equation:

$$4Fe^{3+} + 3BH^{4-} + 9H_2O \rightarrow 4Fe(s) + 3H_2BO_3^- + 12H^+ + 6H_2$$

This method poses two major drawbacks: (1) use of highly acidic and very hygroscopic ferric chloride salt and (2) excessive chloride levels in final products. To overcome these drawbacks, the second-generation methods have been used, which incorporate the use of iron sulfates in place of iron chloride. In this process, the stoichiometric excess of borohydride to iron is 3.6, which is less than the chloride method, where it is 7.4 times more approximately [49].

Besides this, nZVI has also been reported for synthesis by the reduction of goethite (FeOOH) with heat and H_2 and decomposition of iron pentacarbonyl [$Fe(CO)_5$] in organic solvents under inert atmosphere [51, 52]. The nZVI has also been synthesized using the vacuum sputtering or chemical vapor deposition techniques in large quantities [53, 54]. In another approach, researchers [55, 56] have used simple, low cost, and quick method for the production of nZVI nanoparticles using electrolysis. In this approach, Fe^{2+} salt solution has been used with a conductive substrate and direct current has been passed and the nanoparticle synthesized were deposited on the cathode, further collected using ultrasonication.

In addition, nZVI nanoparticles have also played an important role in water purification. The nZVI nanoparticles have been used for both in situ and ex situ treatment of contaminated groundwater. These particles work both as an adsorbent and a reducing agent, thereby degrading toxic organic pollutants into less toxic simple carbon compounds and heavy metals, which agglomerate easily and stick to the soil surface. For in situ treatment, nZVI nanoparticles have been injected directly into the source of contaminated groundwater in the form of slurry, and for ex situ treatment, membranes impregnated with nZVI nanoparticles have been used.

The first application of zero-valent iron (nZVI) nanoparticles has been for the remediation of chlorohydrocarbons in contaminated water [57]. During this reaction, nZVI nanoparticles follow the mechanism of corrosion, wherein the oxidation of iron takes place, releasing electrons, which then reduces organic pollutant through dechlorination.

$$Fe^0 + RCl + H^+ \rightarrow Fe^{2+} + RH + Cl^-$$

The nZVI nanoparticles have also been used to convert from the carcinogenic hexavalent form of Cr to nontoxic Cr^{3+} form, which can readily precipitate as $Cr(OH)_3$ or as the solid solution $Fe_xCr_{1-x}(OH)_3$ [58].

$$Fe^{2+} + CrO_4^{2-}\, 4H_2O \rightarrow (Fe_x, Cr_{1-x})(OH)_3 + 5OH^-$$

Apart from this, nZVI nanoparticles have helped to control the pesticides action such as hexachlorobenzene, hexachlorocyclohexane, chlordane, and dieldrin [59]. Another important application of nZVI nanoparticles has been the removal of arsenic (As) from groundwater. Arsenic has been known to accumulate in liver, kidney, lung, and skin tissues by ingestion of arsenic in drinking water causing a number of diseases. There have been increased reports of arsenic poisonings from countries such as Bangladesh, India, Argentina, Taiwan, and Mexico [60]. It has been shown that nZVI nanoparticles can effectively remove both As (III) and As (V) from water, and the removal efficiency of arsenic depends on the surface area and increased with time (as the surface area of iron increases with time due to corrosion). Also, it has been shown that with proper design, it has been possible to remove arsenic to levels below 5 ppb [61]. Recently zero-valent iron (nZVI) nanoparticles supported on activated carbon have been used as a point of use to remove arsenic [62]. These particles have been successfully used to remove other heavy metals such as Co^{2+}, Pb^{2+}, Ni^{2+}, etc [63]. Using these nanoparticles, dyes based on azo, anthraquinone, and triphenylmethane groups have successfully been degraded for purification of water [64]. nZVI nanoparticles have also been shown to be an effective bactericide against *E. coli*, *B. subtilis*, and *Pseudomonas fluorescens* [65]. These particles have effectively inactivated virus MS2 coliphage and fungus *Aspergillus versicolor* [66]. Hence, nZVI nanoparticles work very effectively for water purification.

Another type of iron nanoparticle that has the potential to be used for water treatment is superparamagnetic iron oxide nanoparticles [67]. The basic principle of water purification by these magnetic Fe_3O_4 nanoparticles is chemical or physical adsorption. Therefore, it can be used for organic pollutant as well as heavy metal ions removal from polluted water. These particles also have the advantage that they have low toxicity and chemical inert, which can increase their applicability for water treatment.

For the effective use of magnetic Fe_3O_4 nanoparticles, the synthesis of these nanoparticles plays an important role. The size distribution, morphology, magnetic properties, and surface chemistry of magnetic Fe_3O_4 nanoparticles depend on the preparation methods. A number of methods have been reported for their synthesis such as coprecipitation, solvothermal, hydrothermal synthesis, microemulsion, and sonochemical method. The simplest and highly efficient method to obtain magnetic Fe_3O_4 particles has been coprecipitation. In this method, a stoichiometric mixture of ferrous and ferric precursors is used as an iron source under alkaline conditions to obtain superparamagnetic nanoparticles [67]. The composition, shape, and size of the magnetic nanoparticles depend on many parameters such as type of salts used (e.g., chlorides, sulfates, nitrates), ratio of Fe^{2+}/Fe^{3+}, reaction temperature, types of stabilizing agent, pH value, and ionic strength of the reaction mixture

[68]. The hydrothermal synthesis has been used for the synthesis of large size nanoparticles with high crystallinity and controlled morphology [69].

According to the sonochemical approach, synthesis of magnetic nanoparticles with an average size of 10 nm has been reported using iron(II)acetate as the iron source in water and irradiated reaction mixture with a high-intensity ultrasonic probe (20 kHz) under 1.5 atm at 25 °C [70]. In addition, oil-in-water emulsions have been used for the synthesis of magnetic nanoparticles; in this approach, cyclohexane has been used as an oil phase and polyoxyethyleneisooctyl ether phosphate (NP$_5$) and nonoxynol-9 phosphate (NP$_9$) as a surfactant. FeSO$_4$/Fe(NO$_3$)$_3$ salt mixture has been taken as an iron source. The resulting nanoparticles have been assumed to possess needle, rods, and hollow spheres shape in morphology. The major advantage of this approach has been that one can control the size of the nanoparticles by controlling the aqueous phase [71].

Nowadays, these magnetic Fe$_3$O$_4$ nanoparticles have been mostly used for water purification because it is easily separated from water after purification. Additionally, their low cost, strong adsorption capacity, easy separation and enhanced stability, magnetic Fe$_3$O$_4$ nanoparticles have emerged as a promising candidate for large-scale wastewater treatment. The Fe$_3$O$_4$ nanoparticles have been mainly applied as a nanoabsorbent for removing pollutants. There has been a great increase in pollution of water by heavy metal due to rapid industrialization and ability to bioaccumulate even at low concentration. Also, these metals have toxic effects on plants, animals, and human beings; therefore, their removal from water is of utmost significance for improving water quality. Due to their magnetic properties, Fe$_3$O$_4$ nanoparticles have been easily separated by applying magnetic field and thus easily used for wastewater treatment. The small size of these particles allows diffusion of metal ions from solution onto the active sites of the adsorbent surface, thereby removing the heavy metals from water.

It has been reported that these particles can absorb 36.0 mg g^{-1} of Pb (II) ions, which is comparably higher than other low cost absorbents [72]. The efficiency of these particles has been enhanced by functionalizing them with chelating ligands. The heavy metal after chelation has successfully been removed with the help of external magnetic field; for example, the iron oxide particle modified with 1,6-hexadiamine has been effectively used for the removal of Cu^{2+} ions from water. This system shows good stability and has adsorption capacity of 25 mg g^{-1} [73]. In another report, magnetic nanoparticles encapsulated in carbon have been used for the removal of Cu^{2+} and Cd^{2+}, and using this system, up to 95% removal of Cu^{2+} and Cd^{2+} has been achieved [74]. By modifying magnetic nanoparticles with gum Arabic, it has been shown that the absorption capacity of the particles has been increased. It has been shown that after modification, particle absorption capacity increases from 17.6 mg g^{-1} for the unmodified surface to 38.5 mg g^{-1} for modified surfaces for copper ions. The particle modified with biopolymer Chitosan shows enhanced absorption capacity for Pb^{2+}, Cu^{2+}, and Cd^{2+} [75, 76]. A number of mechanisms have been proposed for the adsorption of pollutants from wastewater such as electrostatic interaction, surface sites binding, and magnetic selective adsorption.

Modified magnetic nanoparticles have also been used for the detection and removal of pesticides from wastewater. Magnetic nanoparticles modified with acetylcholinesterase have been shown to effectively detect pesticide chlorothalonil at very low levels [77]. The particles modified with octadecyltrichlorosilane have been shown to be effective for the removal of phosphorous pesticides from wastewater [78]. The hollow nanospheres of Fe_3O_4 have been shown to be effective in the removal of red dye from water with maximum adsorption capacity of 90 mg g^{-1} [79]. These magnetic particles have also been used for the treatment of wastewater discharged from oil refinery [80]. Similar to heavy metal removal, magnetic nanoparticles have also been used for the removal of organic pollutants from wastewater. In this case also, organic contaminants are first adsorbed via surface exchange reactions until all surface functional sites are fully occupied; after that contaminants diffuse into the adsorbent for further interactions with functional groups [81]. These properties of magnetic particles can be used for the effective removal of organic pollutants such as polyaromatic hydrocarbons [82].

Further, the nanocomposite of Fe_3O_4 with MnO_2 has been used for the removal of As (V) from wastewater [83]. Similarly, the nanocomposite of Fe_3O_4 with TiO_2 has been used for photocatalytic degradation of methyl orange dye [84]. These particles have been effectively used as an antibacterial agent against *Staphylococcus aureus* [85], and it has been shown that the antibacterial properties of these particles could be enhanced further by the synthesis of nanocomposite with Ag. The minimum inhibitory concentrations of these nanocomposites for *E-coli* and *S. aureus* have been found to be 15.625 and 31.25 mgL^{-1} [86].

Another methodology used for water purification is the use of titanium oxide (TiO$_2$) nanoparticles. The nanosized titanium oxides have outstanding chemical stability, high refractive index, and show enhanced photocatalytic ability, which makes them an ideal candidate for large-scale water treatment. These particles have the tendency to remove heavy metal present in wastewater, and their photocatalytic abilities have been used to degrade organic pollutants in wastewater. Further, these particles show enhanced antibacterial properties due to their photocatalytic abilities, thereby finding application in the removal of bacteria from contaminated water.

There are a number of methods illustrated in literature for the synthesis of TiO$_2$ such as sol-gel method, micelle and inverse micelle method, hydrothermal method, solvothermal method, microwave method, chemical and physical vapor deposition, and sonochemical method [87]. In sol-gel method, the titanium precursor has been hydrolyzed to synthesize the TiO$_2$ nanomaterial; generally titanium (IV) alkoxide has been used as the titanium precursor and subjected to acid-catalyzed hydrolysis followed by condensation to synthesize TiO$_2$ nanomaterial [88]. It has been shown that the shape and size of nanoparticles can be controlled by varying the reaction parameters [89]. Further, TiO$_2$ nanoparticles have also been used for synthesis using reversed microemulsion system of cyclohexane, poly(oxyethylene)$_5$nonyle phenol ether, and poly(oxyethylene)$_9$nonylephenol ether with TiCl$_4$ solution as titanium source and ammonia as a precipitating agent. These amorphous particles have further been converted into anatase and rutile phase by heating them at temperatures from 200 °C to 750 °C and above 750 °C, respectively [90].

Synthesis of TiO$_2$ nanoparticles under room temperature has been reported in reverse micelles system of NP-5 (Igepal CO-520)-cyclohexane, where titanium tetrabutoxide has been hydrolyzed using various acids [91]. TiO$_2$ nanoparticles have also been synthesized using hydrothermal reaction of titanium alkoxide in an acidic ethanol–water solution, and particles synthesized are mainly anatase phase with particle size ranging from 7 nm to 25 nm [92]. In comparison to the hydrothermal method, better control over size and morphology of nanoparticles can be achieved using the solvothermal method. In this method, titanium(IV) isopropoxide (TTIP) has been used as a titanium source and has been mixed with toluene in the weight ratio of 1–3:10 and heated at 250 °C for 3 h [93]. In another report, TiO_2 nanoparticles have been synthesized using the solvothermal approach by controlling the hydrolysis of $Ti(OC_4H_9)_4$ with linoleic acid [94].

Another method for the synthesis of TiO_2 nanoparticles has been the chemical vapor deposition technique. In this approach, TTIP (titanium precursor) has been pyrolysised under helium or oxygen atmosphere to obtain nanoparticles of size below 10 nm [95]. Similarly, TiO_2 nanoparticles have also been synthesized using physical vapor deposition techniques [96]. In the sonochemical method for the synthesis of TiO_2 nanoparticles with enhanced photoactivity, hydrolysis of titanium(IV) isoproproxide has been carried in water or in a 1:1 $EtOH-H_2O$ solution under ultrasonic radiation [97]. The synthesized particle contains both anatase and brookite phases and has high photoactivity. Microwave irradiation has also been used for the synthesis of colloidal TiO_2 nanoparticles suspension, in comparatively less time than the conventional heating method [98, 99].

These TiO_2 nanoparticles have been used for water purification systems in large scale due to their specific properties. TiO_2 nanoparticles have more specific surface area 185.5 m^2g^{-1} in comparison to bulk, which has m^2g^{-1}, due to which they can function as efficient adsorbent for heavy metals. It has been shown that these TiO_2 nanoparticles have been able to remove multiple heavy metals such as Zn, Cd, Pb, Ni, and Cu at pH 8 from wastewater [100]. Removal of heavy metals using TiO_2 nanoparticles follows the modified first-order adsorption kinetics. In another report, TiO_2 nanoparticles having size from 10 nm to 50 nm and surface area of 208 m^2g^{-1} showed the adsorption capacity of 15.3 m^2g^{-1} and 7.9 m^2g^{-1} for Zn^{2+} and Cd^{2+} ions, respectively [101]. Modifying TiO_2 nanoparticles with chelating ligand leads to enhanced metal adsorption capabilities; TiO_2 nanoparticle modified with thiolactic acid, cysteine, and alanine show selective and enhanced adsorption of Pb^{2+} and Cu^{2+} ions [102]. Similarly, TiO_2 nanoparticles modified with arginine have been used for the removal of mercury from water, and these modified particles remove about 60% of initial mercury present in the water [103]. TiO$_2$ nanoparticles have also been used for an effective removal of arsenic from wastewater [104].

Dye-containing wastewater increasingly becomes a major problem in both developed and developing countries. Due to the photocatalytic activity of TiO$_2$ nanoparticles, they have been effectively used for degradation of these dyes in wastewater for purifying the polluted water. These particles have been used for degradation of commercial cationic blue GRL dye, and the degradation follows the pseudo-first-order kinetics [105]. Other types of dyes that have been degraded using TiO$_2$ nanoparticles include diazo dye, azo dye, and anthraquinone dye [106]. These nanoparticles have also been used to degrade toxic polychlorinated biphenyls (PCBs) present in wastewaters. The planar PCBs have been more effectively degraded than nonplanar ones using TiO$_2$ nanoparticles [107].

The presence of pesticides in water from agricultural runoff has been another major concern that affects the quality of potable water. TiO$_2$ nanoparticles have been effectively used for the degradation of pesticides present in water. Using TiO$_2$ nanoparticles, it has been shown that herbicide erioglaucine has been degraded very effectively; the degradation follows second-order kinetics [108]. Similarly, s-triazine herbicides and organophosphorus insecticides have been degraded using the aqueous suspensions of TiO$_2$ nanoparticles [109].

The photocatalytic properties of TiO$_2$ nanoparticles have been used for the inactivation of bacteria present in water. TiO$_2$ nanoparticles have been efficient in killing both Gram-negative and Gram-positive bacteria. But due to their ability to form spores, Gram-positive bacteria are less sensitive to TiO$_2$ nanoparticles [110]. The effective concentration of TiO$_2$ nanoparticles for the inactivation of bacteria depends on the size of particles and the intensity of the light used. These nanoparticles have been able to successfully inhibit the growth of viruses such as *poliovirus 1*, *hepatitis B virus*, *herpes simplex virus*, and *MS2 bacteriophage* [111]. The antibacterial properties of TiO$_2$ have been enhanced by doping them with Ag metal [112]. TiO$_2$ nanomaterials have been able to activate the ROS production in bacteria, especially hydroxyl free radicals and peroxide formed under UV irradiation using the oxidative and reductive pathways, thereby inhibiting bacterial growth [112]. This shows that TiO_2 nanoparticles have the ability to remove organic, inorganic pollutants, and antibacterial properties and hence have the potential for commercial application for wastewater treatment.

Conclusion

Nanoparticles have been playing a major role in water purification nowadays. Among all nanoparticles available in literature, a few of them show the tremendous role in water purification, such as Ag, zero-valent iron, Fe_3O_4, and TiO_2 nanoparticles. Ag nanoparticles have generally been used for the removal of pathogens from water and heavy metals from many years. Recently, nZVI nanoparticles have been used mainly for the degrading toxic organic pollutants and removal of heavy metals. They have been found to be useful for the removal of As (III) and As (V) from water, which is a significant work in water purification. Fe_3O_4 nanoparticles have been used as nanoabsorbent for the removal of pollutants, and these nanoparticles have been easily removed by applying magnetic field. Further, TiO_2 nanoparticles have effectively been used for the degradation of pesticides present in water, and due to their catalytic property, they have been used for the inactivation of bacteria present in water. Hence, nanoparticles have tremendous application in water purifications. Using these nanoparticles, we can provide clean drinking water to growing population.

References

1. UNDP. (2011). *Human Development Report 2011*, UNDP, New York.
2. Keast, G., and Johnston, R. (2008). *UNICEF Handbook on Water Quality*, UNICEF, New York.
3. WHO/UNICEF. (2005). *Water for Life: Making it Happen*, World Health Organization and UNICEF, Geneva, ISBN 9241562935.
4. Nieuwenhuijsen, M. J. (2000). Chlorination disinfection byproducts in water and their association with adverse reproductive outcomes: A review, *Occup. Environ. Med.*, **57**, pp. 73–85.
5. Boorman, G. A. (1999). Drinking water disinfection byproducts: Review and approach to toxicity evaluation, *Environ. Health Persp.*, **107** (suppl 1), pp. 207–217.

6. Hrudey, S. E., and Charrois, J. W. A. (2012). *Disinfection By-Products (DBPs) as a Public Health Issue in Disinfection By-Products and Human Health*, IWA Publishing, London, pp. 1–10.
7. Woo, Y.-T., Lai, D., McLain, J. L., Manibusan, M. K., and Dellarco, V. (2002). Use of mechanism based structure-activity relationships analysis in carcinogenic potential ranking for drinking water disinfection by-products, *Environ. Health Persp.*, **110**, pp. 75–88.
8. Hillie, T., and Hlophe, M. (2007). Nanotechnology and the challenge of clean water, *Nat. Nanotechnol.*, **2**, pp. 663–664.
9. Brame, J., Li, Q., and Pedro, J. J. (2011). Alvarez nanotechnology-enabled water treatment and reuse: Emerging opportunities and challenges for developing countries, *Trends Food Sci. Technol.*, **22**, pp. 618–624.
10. Khin, M. M., Nair, A. S., Babu, V. J., Murugan, R., and Ramakrishna, S. (2012). A review on nanomaterials for environmental remediation, *Energy Environ. Sci.*, **5**, pp. 8075–8109.
11. Castellano, J. J., Shafii, S. M., Ko, F., Donate, G., Wright, T. E., Mannari, R. J., Payne, W. G., Smith, D. J., and Robson, M. C. (2007). Comparative evaluation of Ag-containing antimicrobial dressings and drugs, *Int. Wound J.*, **4**, pp. 14–22.
12. Vaidyanathan, R., Kalishwaralal, K., Gopalram, S., and Gurunathan, S. (2009). Nano-Ag—The burgeoning therapeutic molecule and its green synthesis, *Biotechnol. Adv.*, **27**, pp. 24–37.
13. Kim, J. S., Kuk, E., Yu, K. N., Kim, J. H., Park, S. J., Lee, H. J., Jeong, D. H., and Cho, M. H. (2007). Antimicrobial effects of Ag nanoparticles, *Nanomed. Nanotechnol. Biol. Med.*, **3**, pp. 95–101.
14. Chen, X., and Schluesener, H. J. (2008). Nano-Ag: A nanoproduct in medical application, *Toxicol. Lett.*, **176**, pp. 1–12.
15. Kim, K. J., Sung, W. S., Suh, B. K., Moon, S. K., Choi, J. S., Kim, J. G., and Lee, D. G. (2009). Antifungal activity and mode of action of Ag nanoparticles on Candida albicans, *Biometals*, **22**, pp. 235–242.
16. Sun, R. W., Chen, R., Chung, N. P., Ho, C. M., Lin, C. L., and Che, C. M. (2005). Ag nanoparticles fabricated in Hepes buffer exhibit cytoprotective activities toward HIV-1 infected cells, *Chem. Commun.*, **40**, pp. 5059–5061.
17. Galdiero, S., Falanga, A., Vitiello, M., Cantisani, M., Marra, V., and Galdiero, M. (2011). Ag nanoparticles as potential antiviral agents, *Molecules*, **16**, pp. 8894–8918.
18. Wiley, B., Sun, Y., Mayers, B., and Xia, Y. (2005). Shape-controlled synthesis of metal nanostructures: The case of Ag, *Chem. Eur. J.*, **11**, pp. 454–463.

19. Tao, A., Sinsermsuksakul, P., and Yang, P. (2006). Polyhedral Ag nanocrystals with distinct scattering signatures, *Angew. Chem. Int. Ed.*, **45**, pp. 4597–4601.
20. Pillai, Z. S., and Kamat, P. V. (2004). What factors control the size and shape of Ag nanoparticles in the citrate ion reduction method? *J. Phys. Chem. B*, **108**, pp. 945–951.
21. Khatoon, U. T., Rao, K. V., Rao, J. V. R., and Aparna, Y. (2011). Synthesis and characterization of Ag nanoparticles by chemical reduction method, *Proc. Inter. Confer. Nanosci., Eng. Tech (ICONSET)*, pp. 97–99.
22. Hong, H.-K., Gong, M.-S., and Park, C.-K. (2009). A facile preparation of Ag nanocolloids by hydrogen reduction of an Ag alkylcarbamate complex, *Bull. Korean Chem. Soc.*, **30**, pp. 2269–2274.
23. Kapoor, S., Lawless, D., Kennepohl, P., Meisel, D., and Serpone, N. (1994). Reduction and aggregation of Ag ions in aqueous gelatin solutions, *Langmuir*, **10**, pp. 3018–3022.
24. Schneider, S., Halbig, P., Grau, H., and Nickel, U. (2008). Reproducible preparation of Ag sols with uniform particle size for application in surface-enhanced Raman spectroscopy, *Photochem. Photobiol.*, **60**, pp. 605–610.
25. Monteiro, D. R., Gorup, L. F., Takamiya, A. S., Colla, A., Filho, R., de Camargo, E. R., and Barbosa, D. B. (2009). The growing importance of materials that prevent microbial adhesion: Antimicrobial effect of medical devices containing Ag, *Inter. J. Antimicro. Agents*, **34**, pp. 103–110.
26. Kim, S., Yoo, B., Chun, K., Kang, W., Choo, J., Gong, M., and Joo, S. (2005). Catalytic effect of laser ablated Ni nanoparticles in the oxidative addition reaction for a coupling reagent of benzylchloride and bromoacetonitrile, *J. Mol. Catal. A: Chem.*, **226**, pp. 231–234.
27. Kawasaki, M., and Nishimura, N. (2006). 1064-nm laser fragmentation of thin Au and Ag flakes in acetone for highly productive pathway to stable metal nanoparticles, *Appl. Surf. Sci.*, **253**, pp. 2208–2216.
28. Link, S., Burda, C., Nikoobakht, B., and El-Sayed, M. (2000). Laser-induced shape changes of colloidal gold nanorods using femtosecond and nanosecond laser pulses, *J. Phys. Chem. B*, **104**, pp. 6152–6163.
29. Tarasenko, N., Butsen, A., Nevar, E., and Savastenko, N. (2006). Synthesis of nanosized particles during laser ablation of gold in water, *Appl. Surf. Sci.*, **252**, pp. 4439–4444.
30. Ahmad, A., Mukherjee, P., Senapati, S., Mandal, D., Khan, M. I., Kumar, R., and Sastry, M. (2003). Extracellular biosynthesis of Ag nanoparticles

using the fungus Fusariumoxysporum, *Colloid. Surface. B*, **28**, pp. 313–318.
31. Begum, N. A., Mondal, S., Basu, S., Laskar, R. A., and Mandal, D. (2009). Biogenic synthesis of Au and Ag nanoparticles using aqueous solutions of Black Tea leaf extracts, *Colloid. Surface. B*, **71**, pp. 113–118.
32. Kalishwaralal, K., Deepak, V., Ramkumarpandian, S., Bilal, M., and Sangiliyandi, G. (2008). Biosynthesis of Ag nanocrystals by *Bacillus licheniformis*, *Colloid. Surface. B*, **65**, pp. 150–153.
33. Saifuddin, N., Wong, C. W., and Nur Yasumira, A. A. (2009). Rapid biosynthesis of Ag nanoparticles using culture supernatant of bacteria with microwave irradiation, *E-J. Chem.*, **6**, pp. 61–70.
34. Shahverdi, A. R., Minaeian, S., Shahverdi, H. R., Jamalifar, H., and Nohi, A. (2007). Rapid synthesis of Ag nanoparticles using culture supernatants of Enterobacteria: A novel biological approach, *Process Biochem.*, **42**, pp. 919–923.
35. Kumar, S. A., Majid Kazemian, A., Gosavi, S. W., Sulabha, K. K., Renu, P., Ahmad A., and Khan, M. I. (2007). Nitrate reductase-mediated synthesis of Ag nanoparticles from AgNO$_3$, *Biotechnol. Lett.*, **29**, pp. 439–445.
36. Ingle, A., Gade, A., Pierrat, S., Sönnichsen, C., and Mahendra, R. (2008). Mycosynthesis of Ag nanoparticles using the fungus *Fusarium acuminatum* and its activity against some human pathogenic bacteria, *Curr. Nanosci.*, **4**, pp. 141–144.
37. Vigneshwaran, N., Kathe, A. A., Varadarajan, P. V., Nachane, R. P., and Balasubramanya, R. (2006). Biomimetics of Ag nanoparticles by white rot fungus, Phaenerochaetechrysosporium, *Colloid. Surface. B*, **53**, pp. 55–59.
38. Vigneshwaran, N., Ashtaputre, N. M., Varadarajan, P. V., Nachane, R. P., Paralikar, K. M., and Balasubramanya, R. (2007). Biological synthesis of Ag nanoparticles using the fungus Aspergillusflavus, *Mater. Lett.*, **61**, pp. 1413–1418.
39. Begum, N. A., Mondal, S., Basu, S., Laskar, R. A., and Mandal, D. (2009). Biogenic synthesis of Au and Ag nanoparticles using aqueous solutions of Black Tea leaf extracts. *Colloid. Surface. B,* **71**, pp. 113–118.
40. Sondi, I., and Salopek-Sondi, B. (2004). Silver nanoparticles as antimicrobial agent: A case study on *E. coli* as a model for gram-negative bacteria, *J. Coll. Inter. Sci.*, **275**, pp. 177–182.
41. Feng, Q. L., Wu, J., Chen, G. Q., Cui, F. Z., Kim, T. N., and Kim, J. O. (2000). A mechanistic study of the antibacterial effect of Ag ions on *Escherichia coli* and *Staphylococcus aureus*, *J. Biomed. Mat. Res.*, **52**, pp. 662–668.

42. Maynard, A. D. (2007). Nanotechnology: Toxicological issues and environmental safety and environmental safety, in *Project on Emerging Nanotechnologies*, pp. 1–14. Woodrow Wilson International Center for Scholars, Washington, DC.

43. Nair, A. S., and Pradeep, T.Extraction of chlorpyrifos and malathion from water by metal nanoparticles, *J. Nanosci. Nanotechnol.*, **7**, pp. 1871–1877.

44. Bootharaju, M. S., and Pradeep. T. (2012). Understanding the degradation pathway of the pesticide, chlorpyrifos by noble metal nanoparticles, *Langmuir*, **28**, pp. 2671–2679.

45. Henglein, A. (1998). Colloidal Ag nanoparticles: Photochemical preparation and interaction with O_2, CCl_4, and some metal ions, *Chem. Mater.*, **10**, pp. 444–450

46. Sumesh, E., Bootharaju, M. S., Anshup, S., and Pradeep, T. (2011). A practical Ag nanoparticle-based adsorbent for the removal of Hg^{2+} from water, *J. Hazard. Mater.*, **189**, pp. 450–457

47. Lin, S., Huang, R., Cheng, Y., Liu, J., Lau, B. L. T., and Wiesner, M. R., (2012). Ag nanoparticle-alginate composite beads for point-of-use drinking water disinfection, *Water Res.*, doi: 10.1016/j.watres.2012.09.005.

48. Saifuddin, N., Nian, C. Y., Zhan, L. W., and Ning, K. X. (2011). Chitosan-Ag nanoparticles composite as point-of-use drinking water filtration system for household to remove pesticides in water, *Asian J. Biochem.*, **6**, pp. 142–159.

49. Matheson, L. J., and Tratnyek, P. G. (1994). Reductive dehalogenation of chlorinated methanes by iron metal. *Environ. Sci. Technol.*, **28**, pp. 2045–2053.

50. Wang, C.-B., and Zhang, W.-X. (1997). Synthesizing nanoscale iron particles for rapid and complete dechlorination of TCE and PCBs, *Environ. Sci. Technol.*, **31**, pp. 2154–2156.

51. Nurmi, J. T., Tratnyek, P. G., Sarathy V., Baer, D. R., Amonette, J. E., Pecher, K., Wang, C., Linehan, J. C., Matson, D. W., Penn, R. L., and Driessen, M. D. (2005). Characterization and properties of metallic iron nanoparticles: Spectroscopy, electrochemistry, and kinetics, *Environ. Sci. Technol.*, 39:1221–1230.

52. Karlsson, M. N. A., Deppert, K., Wacaser, B. A. Karlsson, L.S., and Malm, J.-O. (2005). Size-controlled nanoparticles by thermal cracking of iron pentacarbonyl, *Appl. Phys. A*, **80**, pp. 1579–1583.

53. Kuhn, L. T., Bojesen, A., Timmermann, L., Meedom Nielsen, M., and Mørup, S. (2002). Structural and magnetic properties of core-shell

iron-iron oxide nanoparticles, *J. Phys.-Condens. Mat.*, **14**, pp. 13551–13567.

54. Zaera, F. (1989). A thermal desorption and X-ray photoelectron spectroscopy study of the surface chemistry of iron pentacarbonyl, *J. Vac. Sci. Technol. A*, **7**, pp. 640–645.

55. Hoag, G. E., Collins, J. B., Holcomb, J. L., Hoag, J. R., Nadagouda, M. N., and Varma, R. S. (2009). Degradation of bromothymol blue by 'greener' nanoscale zero-valent iron synthesized using tea polyphenols, *J. Mater. Chem.*, **19**, pp. 8671–8677.

56. Chen, S., Hsu, H., Li, C. (2005). A new method to produce nanoscale iron for nitrate removal, *J. Nanopart. Res.*, **6**, pp. 639–647.

57. Matheson, L. J., and Tratnyek, P. G. (1994). Reductive dehalogenation of chlorinated methanes by iron metal. *Environ. Sci. Technol.*, **28**, pp. 2045–2053.

58. Puls, R. W., Paul, C. J., and Powell, R. M. (1999). The application of in situ permeable reactive (zero-valent iron) barrier technology for the remediation of chromate-contaminated groundwater: A field test, *Appl. Geochem.*, **14**, pp. 989–1000.

59. Elliott, D. W. (2005). Iron nanoparticles: Reactions with lindane and the hexachlorocyclohexanes. Doctoral dissertation, Department of Civil and Environmental Engineering, Lehigh University, Bethlehem, PA.

60. Nikolaidis, N. P., Dobbs, G. M., and Lackovic, J. A. (2003). Arsenic removal by zero-valent iron: Field, laboratory and modeling studies, *Water Res.*, **37**, pp. 1417–1425.

61. Lackovic, J. A., Nikolaidis, N. P., and Dobbs, G. M. (2000). Inorganic arsenic removal by zero-valent iron, *Environ. Eng. Sci.*, **17**, pp. 29–39.

62. Zhua, H., Jia, Y., Wu, X., and Wanga, H. (2009). Removal of arsenic from water by supported nano zero-valent iron on activated carbon, *J. Hazard. Mater.*, **172**, pp. 1591–1596.

63. Noubactep, C., Caré, S., and Crane, R. (2012). Nanoscale metallic iron for environmental remediation: Prospects and limitations, *Water Air Soil Poll.*, **223**, pp. 1363–1382.

64. He, Y., Gao, J.-F., Feng, F.-Q., Liu, C., Peng, Y.-Z., and Wang, S.-Y. (2012). The comparative study on the rapid decolorization of azo, anthraquinone and triphenylmethane dyes by zero-valent iron, *Chem. Eng. J.*, **179**, pp. 8–18.

65. Diao, M., and Yao, M. (2009). Use of zero-valent iron nanoparticles in inactivating microbes, *Water Res.*, **43**, pp. 5243–5251.

66. Kim, J. Y., Lee, C., Love, D. C., Sedlak, D. L., Yoon, J., and Nelson, K. L. (2011). Inactivation of MS2 coliphage by ferrous ion and zero-valent iron nanoparticles, *Environ. Sci. Technol.*, **45**, pp. 6978–6984.
67. Kang, Y. S., Risbud, S., Rabolt, J. F., and Stroeve, P. (1996). Synthesis and characterization of nanometer-size Fe_3O_4 and Fe_2O_3 particles, *Chem. Mater.*, **8**, pp. 2209–2211.
68. Laurent, S., Forge, D., Port, M., Roch, A., Robic, C., Vander Elst, L., and Muller, R. N. (2008). Magnetic iron oxide nanoparticles: Synthesis, stabilization, vectorization, physicochemical characterizations, and biological applications, *Chem. Rev.*, **108**, pp. 2064–2110.
69. Chen, D., and Xu, R. (1998). Hydrothermal synthesis and characterization of nanocrystalline Fe_3O_4 powders, *Mater. Res. Bull.*, **33**, pp. 1015–1021.
70. Vijayakumar, R., Koltypin, Y., Felner, I., and Gedanken, A. (2000). Sonochemical synthesis and characterization of pure nanometer-sized Fe_3O_4 particles, *Mater. Sci. Eng. A*, **286**, pp. 101–105.
71. Zhou, Z. H., Wang, J., Liu, X., and Chan, H. S. O. (2001). Synthesis of Fe_3O_4 nanoparticles from emulsions, *J. Mater. Chem.*, **11**, pp. 1704–1709.
72. Nassar, N. N. (2010). Rapid removal and recovery of Pb(II) from wastewater by magnetic nanoadsorbents, *J. Hazard. Mater.*, **184**, pp. 538–546.
73. Hao, Y. M., Man, C., and Hu, Z. B. (2010). Effective removal of Cu(II) ions from aqueous solution by aminofunctionalized magnetic nanoparticles, *J. Hazard. Mater.*, **184**, pp. 392–399.
74. Bystrzejewski, M., Pyrzyńska, K., Huczko, A., and Lange, H. (2009). Carbon-encapsulated magnetic nanoparticles as separable and mobile sorbents of heavy metal ions from aqueous solutions, *Carbon*, **47**, pp. 1201–1204.
75. Banerjee, S. S., and Chen, D. H. (2007). Fast removal of copper ions by gum arabic modified magnetic nanoadsorbent, *J. Hazard. Mater.*, **147**, pp. 792–799.
76. Liu, X. W., Hu, Q. Y., Fang, Z., Zhang, X. J., and Zhang, B. B. (2008). Magnetic chitosan nanocomposites: A useful recyclable tool for heavy metal ion removal, *Langmuir*, **25**, pp. 3–8.
77. Lawruk, T. S., Lachman, C. E., Jourdan, S. W., Fleeker, J. R., Herzog, D. P., and Rubio, F. M. (1993). Quantification of cyanazine in water and soil by a magnetic particle-based ELISA, *J. Agri. Food Chem.*, **41**, pp. 747–752.

78. Shen, H. Y., Zhu, Y., Wen, X. E., and Zhuang, Y. M. (2007). Preparation of Fe$_3$O$_4$-C$_{18}$ nano-magnetic composite materials and their cleanup properties for organophosphorous pesticides, *Anal. Bioanal. Chem.*, **387**, pp. 2227–2237.
79. Iram, M., Guo, C., Guan, Y. P., Ishfaq, A., and Liu, H. Z. (2010). Adsorption and magnetic removal of neutral red dye from aqueous solution using Fe$_3$O$_4$ hollow nanospheres, *J. Hazard. Mater.*, **181**, pp. 1039–1050.
80. Petrakis, L., and Ahner, P. F. (1976). Use of high gradient magnetic separation techniques for the removal of solids from water effluents, *IEEE T. Magn.*, **12**, pp. 486–487.
81. Hu, J., Shao, D. D., Chen, C. L., Sheng, G. D., Ren, X. M., and Wang, X. K. (2011). Removal of 1-naphthylamine from aqueous solution by multiwall carbon nanotubes/iron oxides/cyclodextrin composite, *J. Hazard. Mater.*, **185**, 463–471.
82. Zhang, S. X., Niu, H. Y., Hu, Z. J., Cai, Y. Q., and Shi, Y. L. (2010). Preparation of carbon coated Fe$_3$O$_4$ nanoparticles and their application for solid-phase extraction of polycyclic aromatic hydrocarbons from environmental water samples, *J. Chromatogr. A*, **1217**, pp. 4757–4764.
83. Zhao, Z., Liu, J., Cui, F., Feng, H., and Zhang, L. (2012). One-pot synthesis of tunable Fe$_3$O$_4$–MnO$_2$ core-shell nanoplates and their applications for water purification, *J. Mater. Chem.*, **22**, pp. 9052–9057.
84. Kojima, T., Gad-Allah, T. A., Kato, S., Satokawa, S. (2011). Photocatalytic activity of magnetically separable TiO$_2$/SiO$_2$/Fe$_3$O$_4$ composite for dye degradation, *J. Chem. Eng. Japan*, **44**, pp. 662–667.
85. Tran, N., Mir, A., Mallik, D., Sinha, A., Nayar, S., and Webster, T. J. (2010). Bactericidal effect of iron oxide nanoparticles on *Staphylococcus aureus*, *Int. J. Nanomed.*, **5**, pp. 277–283.
86. Zhang, X., Niu, H., Yan, J., and Cai, Y. (2011). Immobilizing Ag nanoparticles onto the surface of magnetic silica composite to prepare magnetic disinfectant with enhanced stability and antibacterial activity, *Colloid. Surf. A*, **375**, pp. 186–192.
87. Chen, X., and Mao, S. S. (2007). Titanium dioxide nanomaterials: Synthesis, properties, modifications, and applications, *Chem. Rev.*, **107**, pp. 2891–2959.
88. Bessekhouad, Y., Robert, D., and Weber, J. V. (2003). Synthesis of photocatalytic TiO$_2$ nanoparticles: Optimization of the preparation conditions, *J. Photochem. Photobiol. A*, **157**, pp. 47–53.

89. Sugimoto, T., Okada, K., and Itoh, H. (1997). Synthetic of uniform spindle-type titania particles by the Gel-Sol method, *J. Colloid. Interf. Sci.*, **193**, pp. 140–143.

90. Li, G. L., and Wang, G. H. (1999). Synthesis of nanometer-sized TiO$_2$ particles by a microemulsion method, *Nanostruct. Mater.*, **11**, pp. 663–668.

91. Zhang, D., Qi, L., Ma, J., and Cheng, H. (2002). Formation of crystalline nanosizedtitania in reverse micelles at room temperature, *J. Mater. Chem.*, **12**, pp. 3677–3680.

92. Chae, S. Y., Park, M. K., Lee, S. K., Kim, T. Y., Kim, S. K., and Lee, W. I. (2003). Preparation of size-controlled TiO$_2$ nanoparticles and derivation of optically transparent photocatalytic films, *Chem. Mater.*, **15**, pp. 3326–3331.

93. Kim, C. S., Moon, B. K., Park, J. H., Chung, S. T., and Son, S. M. (2003). Synthesis of nanocrystalline TiO$_2$ in toluene by a solvothermal route, *J. Cryst. Growth*, **254**, pp. 405–410.

94. Li, X. L., Peng, Q., Yi, J. X., Wang, X., and Li, Y. D. (2006). Near monodisperse TiO$_2$ nanoparticles and nanorods, *Chem. Eur. J.*, **12**, pp. 2383–2391.

95. Seifried, S., Winterer, M., and Hahn, H. (2000). Nanocrystallinetitania films and particles by chemical vapor synthesis, *Chem. Vap. Depos.*, **6**, pp. 239–244.

96. Wu, J. M., Shih, H. C., and Wu, W. T. (2005). Electron field emission from single crystalline TiO$_2$ nanowires prepared by thermal evaporation, *Chem. Phys. Lett.*, **413**, pp. 490–494.

97. Yu, J. C., Yu, J., Ho, W., and Zhang, L. (2001). Preparation of highly photocatalytic active nanosized TiO$_2$ particles *via* ultrasonic irradiation, *Chem. Commun.*, pp. 1942–1943.

98. Corradi, A. B., Bondioli, F., Focher, B., Ferrari, A. M., Grippo, C., Mariani, E., and Villa, C. (2005). Conventional and microwave-hydrothermal synthesis of TiO$_2$ nanopowders, *J. Am. Ceram. Soc.*, **88**, pp. 2639–2641.

99. Engates, K. E., and Shipley, H. J. (2011). Adsorption of Pb, Cd, Cu, Zn, and Ni to titanium dioxide nanoparticles: Effect of particle size, solid concentration, and exhaustion, *Environ. Sci. Pollut. Res.*, **18**, pp. 386–395.

100. Liang, P., Shi, T. Q., and Li, J. (2004). Nanometer-size titanium dioxide separation/preconcentration and FAAS determination of trace Zn and Cd in water sample, *Int. J. Environ. Anal. Chem.*, **84**, pp. 315–321.

101. Natalia, M., Laura, S., and Tijana, R. (1999). Removal of heavy metals from aqueous waste streams using surface-modified nanosized TiO$_2$ photocatalysts, *J. Advan. Oxid. Technol.*, **4**, pp. 174–178.

102. Skubal, L. R., and Meshkov, N. K. (2002). Reduction and removal of mercury from water using arginine-modified TiO$_2$. *J. Photochem. Photobiol. A: Chem.*, **148**, pp. 211–214.

103. Hristovski, K., Baumgardner, A., and Westerhoff, P. (2007). Selecting metal oxide nanomaterials for arsenic removal in fixed bed columns: From nanopowders to aggregated nanoparticle media, *J. Hazard. Mater.*, **147**, pp. 265–274.

104. Giwa, A., Nkeonye, P. O., Bello, K. A., and Kolawole, K. A. (2012). Photocatalytic decolourization and degradation of C. I. basic blue 41 using TiO$_2$ nanoparticles, *J. Environ. Prot.*, **3**, pp. 1063–1069.

105. Ahmed, S., Rasul, M. G., Martens, W. N., Brown, R., and Hashib, M. A. (2011). Advances in heterogeneous photocatalytic degradation of phenols and dyes in wastewater: A review, *Water, Air Soil Poll.*, **215**, pp. 3–29.

106. Dasary, S. S. R., Saloni, J., Fletcher, A., Anjaneyulu, Y., and Yu, H. (2010). Photodegradation of selected PCBs in the presence of nano-TiO$_2$ as catalyst and H$_2$O$_2$ as an oxidant, *Int. J. Environ. Res. Public Health*, **7**, pp. 3987–4001.

107. Daneshvar, N., Salari, D., Niaei, A., and Khataee, A. R. (2006). Photocatalytic degradation of the herbicide erioglaucine in the presence of nanosized titanium dioxide: Comparison and modeling of reaction kinetics, *J. Environ. Sci. Health, B*, **41**, pp. 1273–1290.

108. Konstantinou, I., Sakellarides, T., Sakkas, V., and Albanis, T. (2001). Photocatalytic degradation of selected s-triazine herbicides and organophosphorus insecticides over aqueous TiO$_2$ suspensions, *Environ. Sci. Technol.*, **35**, pp. 398–405.

109. Wei, C., Lin, W. Y., Zainal, Z., Williams, N. E., Zhu, K., Kruzic, A. P., Smith, R. L., and Rajeshwar, K. (1994). Bactericidal activity of TiO$_2$ photocatalyst in aqueous media: Toward a solar-assisted water disinfection system, *Environ. Sci. Technol.*, **28**, pp. 934–938.

110. Li, Q., Mahendra, S., Lyon, D. Y., Brunet, L., Liga, M. V., Li, D., and Alvarez, P. J. J. (2008). Antimicrobial nanomaterials for water disinfection and microbial control: Potential applications and implications, *Water Res.*, **42**, pp. 4591–4602.

111. Zhang, H., and Chen, G. (2009). Potent antibacterial activities of Ag/TiO$_2$ nanocomposite powders synthesized by a one-pot sol–gel method, *Environ. Sci. Technol.*, **43**, pp. 2905–2910.

112. Kikuchi, Y., Sunada, K., Iyoda, T., Hashimoto, K., and Fujishima, A. (1997). Photocatalytic bactericidal effect of TiO$_2$ thin films: Dynamic view of the active oxygen species responsible for the effect, *J. Photochem. Photobiol. A. Chem.*, **106**, pp. 51–56.

Chapter 7

Electrochemical Ozone Production for Degradation of Organic Pollutants via Novel Electrodes Coated by Nanocomposite Materials

Mahmoud Abbasi[a] and Ali Reza Soleymani[b]
[a]*Applied Chemistry Department, Faculty of Chemistry,*
University of Mazandaran, Babolsar, Mazandaran, Iran
[b]*Applied Chemistry Department, Faculty of Science,*
Malayer University, Malayer 65174, Iran
mahabbasi79@yahoo.com, m.abbasi@umz.ac.ir

The main scope of this chapter is to overview the recent developments in the electrochemical ozone production, especially in novel anode electrodes coated by nanocomposite materials. But before that briefly, the importance of ozonation process in the wastewater treatment process is expressed, and the oxidation mechanism of ozone molecules operation in the aqueous media is stated. Thereafter, among the different ozone production methods, the electrochemical process will be discussed in detail. The role of novel anode electrodes coated with nanocomposite of Sn–Sb–Ni in the enhancement of ozone production efficiency has been reviewed.

7.1 Introduction

Ozone (O_3), or trioxygen, is a triatomic molecule, consisting of three oxygen atoms. It is an allotrope of oxygen, which is less stable than the diatomic allotrope (O_2), breaking down with a half life of roughly 30 min in the lower atmosphere to normal dioxygen. Ozone is formed from dioxygen by the action of ultraviolet light and also atmospheric electrical discharges. Ozone's odor is sharp, reminiscent of chlorine, and detectable by many people at concentrations of as little as 10 parts per billion in air. Ozone's O_3 formula was determined in 1865. The molecule was later proven to have a bent structure and to be diamagnetic. In standard conditions, ozone is a pale blue gas that condenses at progressively cryogenic temperatures to a dark-blue liquid and finally a violet-black solid. Ozone's instability with regard to more common dioxygen is such that both concentrated gas and liquid ozone may decompose explosively. It is, therefore, used commercially only in low concentrations and produced in situ.

7.2 Ozonation Process in Water and Wastewater Treatment

Biological, chemical, or physical oxidation processes are major steps during water treatment. Biological oxidation is thought to be economically feasible and widely applicable. However, the presence of toxic pollutants in treated water might make the usage of this method impossible. Thermal destruction of chemicals at high temperatures, however effective, is not economically feasible. Chemical oxidation with the usage of several oxidants such as ozone, hydrogen peroxide, chlorine, or chlorine dioxide generally overcomes these difficulties.

Ozone being an environmentally clean reagent with proven efficiency for several processes of technological importance, a renewed interest in its production and application is observed [1–20]. Despite O_3 production via the corona process being rather expensive, the chemical has a number of appealing advantages: It is a very strong oxidant; its decomposition leads to environmentally friendly products (O$_2$); its instability ($t_{1/2}$ = 20–90 min, depending on the environment) requires that it is produced "on spot," thereby reducing transportation and storage expenses. As a result, O$_3$ has found application in fields such as water treatment, combustion of resistant organics, clean-up of effluents, and bleaching of wood pulp [2, 8, 15, 16]. Increasing concern with environmental issues has stimulated research to develop nonaggressive chemicals and improve existing technology (green chemical processes) to minimize or avoid the impact of industrial activity on the environment. Within this context, ozone has received special attention since it is a very powerful oxidant [$E°$=1.51 V (vs. RHE)] not leading to toxic products on decomposition [21–24].

Ozone, due to its high oxidation and disinfection potential, has recently received much attention in water treatment technology. It is applied to improve taste and color as well as to remove the organic and inorganic compounds in water. Several applications of ozone in water treatment technology are subjects of discussion in many books [25–29], and some application instances in the treatment of water supplies are listed in Table 7.1 [30].

Table 7.1 Some instances for ozone application in water treatment

Bacterial disinfection
Viral inactivation
Oxidation of soluble iron and/or manganese
Decomplexing organically bound manganese (oxidation)
Color removal (oxidation)
Algae removal (oxidation)
Oxidation of organics: phenols, detergents, pesticides
Microflocculation of dissolved organics (oxidation)
Oxidation of inorganics: cyanide, sulfide, nitrates
Turbidity or suspended solids removal (oxidation)
Pretreatment for biological processes (oxidation)

7.3 Oxidation Mechanism of Ozonation

The chemistry of ozone in aqueous solution is complex. Molecular ozone can oxidize water impurities via direct, selective reactions or can undergo decomposition via a chain reaction mechanism resulting in the production of free hydroxyl radicals. The chemical properties of ozone depend on the structure of the molecule [31]. Additional free hydroxyl radicals can be produced in the aqueous media from ozone by pH modification or can be introduced by combining ozone either with hydrogen peroxide or UV irradiation from a high pressure mercury lamp [32]. Unlike photocatalysis treatment processes, ozonation process, due to its capability for selective destruction of recalcitrant organics, is used as a pretreatment step before ordinary biological techniques, thus being more efficient for highly contaminated wastewater. Also it is reported that the simultaneous application of ozonation and photocatalysis has the capability for efficient treatment of organically contaminated waters over a wide range of concentrations [33]. The two extreme forms of resonance structures of ozone molecule (Fig. 7.1) can be expressed as follows [34, 35]:

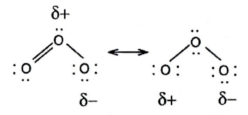

Figure 7.1 Ozone resonance structures.

Due to its structure, molecular ozone can react as a dipole, an electrophilic, or a nucleophilic agent. As a result of its high reactivity, ozone is very unstable in water. The half-life time of molecular ozone varies from a few seconds up to few minutes and depends on pH, water temperature, and concentration of organic and inorganic compounds in water [36, 37]. The decomposition of ozone follows a pseudo first-order kinetic law [34]. Ozone decomposition proceeds through the following five-step chain reaction [37]:

$$O_3 + H_2O \rightarrow 2HO^\bullet + O_2, \quad k = 1.1 \times 10^{-4}\ M^{-1}s^{-1} \quad (7.1)$$

$$O_3 + OH^- \rightarrow O_2^{\bullet-} + HO_2^\bullet, \quad k = 70\ M^{-1}s^{-1} \quad (7.2)$$

$$O_3 + HO^\bullet \rightarrow O_2 + HO_2^\bullet \leftrightarrow O_2^{\bullet-} + H^+ \quad (7.3)$$

$$O_3 + HO_2^{\bullet} \leftrightarrow 2O_2 + HO^{\bullet}, \quad k = 1.6 \times 10^9 \text{ M}^{-1}\text{s}^{-1} \quad (7.4)$$

$$2HO_2^{\bullet} \rightarrow O_2 + H_2O_2 \quad (7.5)$$

The pH value of the solution significantly influences ozone decomposition in water. Basic pH causes an increase of ozone decomposition. At pH < 3, hydroxyl radicals do not influence the decomposition of ozone. For 7< pH<10, the typical half-life time of ozone is from 15 to 25 min. The decomposition of ozone can be significantly lowered in the presence of hydroxyl radical scavengers due to the following reactions [35, 37]:

$$HO^{\bullet} + O_3 \rightarrow O_2 + HO_2^{\bullet}, \quad k = 3.0 \times 10^9 \text{ M}^{-1}\text{s}^{-1} \quad (7.6)$$

$$HO^{\bullet} + HCO_3^- \rightarrow OH^- + HCO_3^{\bullet}, \quad k = 1.5 \times 10^7 \text{ M}^{-1}\text{s}^{-1} \quad (7.7)$$

$$HO^{\bullet} + CO_3^{2-} \rightarrow OH^- + CO_3^{\bullet -}, \quad k = 4.2 \times 10^8 \text{ M}^{-1}\text{s}^{-1} \quad (7.8)$$

$$HO^{\bullet} + H_2PO_4^- \rightarrow OH^- + H_2PO_4^{\bullet}, \quad k < 10^5 \text{ M}^{-1}\text{s}^{-1} \quad (7.9)$$

$$HO^{\bullet} + HPO_4^{2-} \rightarrow OH^- + H_2PO_4^-, \quad k < 10^7 \text{ M}^{-1}\text{s}^{-1} \quad (7.10)$$

In water, there are several compounds that are capable of the initiation, promotion, or inhibition of the radical chain reaction process. The initiators (OH−, H_2O_2/HO_2^-, Fe^{2+}, formate, humic substances) are capable of inducing the formation of superoxide ion $O_2^{\bullet -}$ from an ozone molecule. The promoters [R_2–CH–OH, aryl–(R), formate, humic substances, O_3] are responsible for the regeneration of the $O_2^{\bullet -}$ ion from the hydroxyl radicals. The inhibitors [CH_3–COO−, alkyl–(R), HCO_3^-/CO_3^{2-}, humic substances] are compounds capable of consuming hydroxyl radicals without the regeneration of the superoxide anion [34, 35].

Ozone is used for many different purposes such as disinfection and algae control, taste, odor and color control, oxidation of inorganic pollutants (iron, manganese), oxidation of organic micro- and macropollutants as well as for the improvement of coagulation [34]. There are two main points of oxidant introduction during water treatment: preoxidation, and intermediate oxidation. Generally, preoxidation is applied for the elimination of inorganic compounds, color, taste, odor, turbidity, and suspended solids. During this step, the partial degradation of natural organic matter and inactivation of microorganisms occur as well as the coagulation–flocculation–

decantation step enhancement takes place. Intermediate oxidation has the aim of the degradation of micropollutants, the removal of trihalomethane precursors, and increase of biodegradability [36].

The term advanced oxidation process is defined as the oxidation process that generates hydroxyl radicals in sufficient quantity to affect water treatment. These processes generally use a combination of oxidation agents (ozone, hydrogen peroxide), irradiation (UV, ultrasound), and catalysts as a means of generating hydroxyl radicals. Hydroxyl radical is one of the most reactive free radicals and one of the strongest oxidants [37]:

$$HO^{\bullet} + H^+ + e^- \rightarrow H_2O, \quad E^0 = 2.33V \qquad (7.11)$$

The rate at which hydroxyl radicals react with organic molecules is usually in the order of 10^6–10^9 M^{-1} s^{-1} [38]. The radicals that are formed after hydroxyl radical reaction with organic molecules disproportionate or combine with each other, forming many types of mostly labile intermediates, which react further to produce peroxides, aldehydes, acids, hydrogen peroxide, etc. [39]. The reaction of hydroxyl radicals is not selective. They react rapidly with the primary radical traps, carbonates, bicarbonates, and tert-butanol.

The presented rate constants are only slightly lower than those of hydroxyl radical with organic compounds. This is because hydroxyl radicals, due to their high reactivity, can react with almost all types of organics (ethylenic, lipid, aromatic, aliphatic) and inorganics (anions and cations) [37]. This is the reason why the presence of radical scavengers in water can cause the total inhibition of the free radical chain reaction [34]. The reactivity of hydroxyl radicals with radical scavengers is the main disadvantage of all oxidative degradation processes based on hydroxyl radical reactions.

It is a renowned fact that ozonation can proceed via two routes: direct molecular ozone reactions and/or indirect pathway leading to ozone decomposition and the generation of hydroxyl radicals (OH$^{\bullet}$) [40]. Because hydroxyl radicals produced by ozone decomposition in water are the second strongest oxidizers after fluorine, ozone water is used in a wide range of applications, such as the sterilization of medical instruments, the oxidation and detoxification of wastewater, the deodorization and decolorization of well water, the cleaning of electronic components, and food treatment. For example, the possible applications of ozone water to the control of microbiological safety and the preservation of food quality are well summarized in a review paper [41]. Therefore, the development of equipment that can efficiently produce highly concentrated ozone water and has a long operation lifetime is important for industrial use. Figure 7.2 shows the complete mechanism of ozonation.

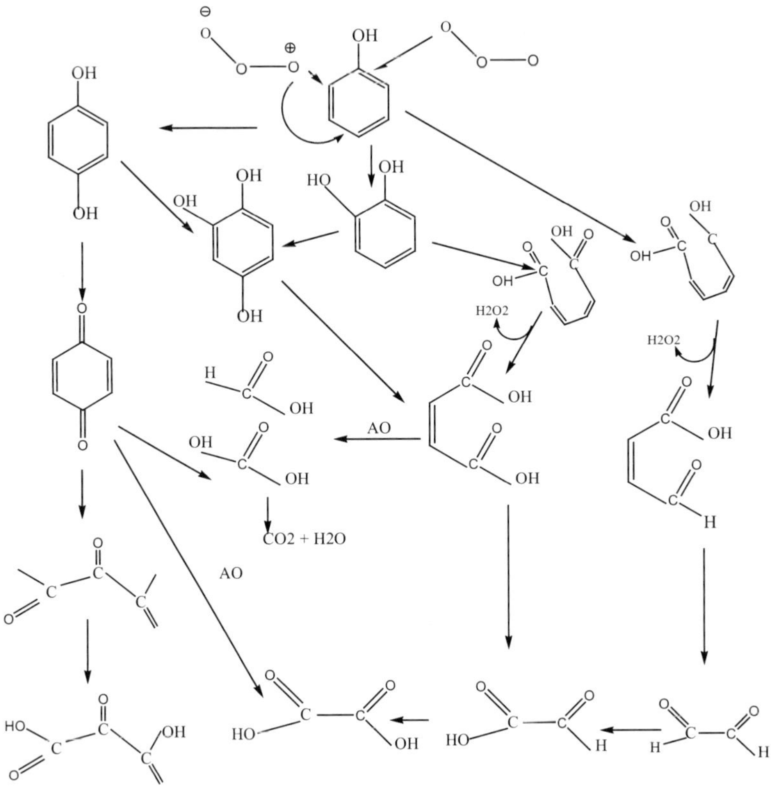

Figure 7.2 Mechanism of ozonation.

7.4 Ozone Production Methods

In nature, ozone often forms under conditions where O_2 will not react [43]. Ozone used in industry is measured in ppm (parts per million), ppb (parts per billion), μgm^{-3}, mgh^{-1}, or weight percent. The regime of applied concentrations ranges from 1% to 5% in air and from 6% to 14% in oxygen for older generation methods. New electrolytic methods can achieve up to 20–30% dissolved ozone concentrations in output water. New electrolytic methods can achieve higher purity and dissolution through using water as the source of ozone production. In the following sections, most important applied methods in the production of ozone will be mentioned in concise, and among them the electrochemical method will be described in more detail.

7.4.1 Corona Discharge Method

The corona discharge method is the most common type of ozone-generating technique for most industrial and personal uses. While variations of the "hot spark" coronal discharge method of ozone production exist, including medical grade and industrial grade ozone generators, these units usually work by means of a corona discharge tube. They are typically cost-effective and do not require an oxygen source other than the ambient air to produce ozone concentrations of 3–6%.

$$\frac{3}{2}O_2 \rightarrow O_3 \qquad \Delta H^\circ_{298} = 34.1 \text{ kcal} \qquad (7.12)$$

Fluctuations in ambient air, due to weather or other environmental conditions, cause variability in ozone production. However, they also produce nitrogen oxides as a by-product [44]. Use of an air dryer can reduce or eliminate nitric acid formation by removing water vapor and increase ozone production. Use of an oxygen concentrator can further increase the ozone production and further reduce the risk of nitric acid formation by removing not only the water vapor, but also the bulk of the nitrogen.

The dominant commercial ozone technology is Cold Corona Discharge (CCD). This technology has a number of disadvantages, including the following:

- Low concentration of O_3 in the output, in the range of 2–12% by volume.
- CCD generators require cold, dry, and pure O_2; if air is employed, the O_3 concentration and production efficiency are significantly reduced and nitrous oxides produced (Seidel, 2004).

- CCD generates only gas phase O_3, which is difficult to dissolve for aqueous applications (Seidel, 2004).
- CCD requires high voltage (kV range) power supplies.

7.4.2 Photochemical Process

UV ozone generators, or vacuum-ultraviolet (VUV) ozone generators, employ a light source that generates a narrow-band ultraviolet light, a subset of that produced by the sun. The sun's UV sustains the ozone layer in the stratosphere of earth [45].

$$O_2 \xrightarrow{hv} 2O^{\bullet} \tag{7.13}$$

$$O_2 + O^{\bullet} \rightarrow O_3 \tag{7.14}$$

While standard UV ozone generators tend to be less expensive, they usually produce ozone with a concentration of about 0.5% or lower. Another disadvantage of this method is that it requires the air (oxygen) to be exposed to the UV source for a longer amount of time, and any gas that is not exposed to the UV source will not be treated. This makes UV generators impractical for use in situations that deal with rapidly moving air or water streams (in-duct air sterilization, for example). Production of ozone is one of the potential dangers of ultraviolet germicidal irradiation. VUV ozone generators are used in swimming pool and spa applications ranging to millions of gallons of water. VUV ozone generators, unlike corona discharge generators, do not produce harmful nitrogen by-products, and also unlike corona discharge systems, VUV ozone generators work extremely well in humid air environments. There is also not normally a need for expensive off-gas mechanisms and no need for air driers or oxygen concentrators, which require extra costs and maintenance.

7.4.3 Cold Plasma

In the cold plasma method, pure oxygen gas is exposed to plasma created by dielectric barrier discharge. The diatomic oxygen is split into single atoms, which then recombine in triplets to form ozone. Cold plasma machines utilize pure oxygen as the input source and produce a maximum concentration of about 5% ozone. They produce far greater quantities of ozone in a given space of time compared to ultraviolet production. However, because cold plasma ozone generators are very expensive, they are found less frequently than the previous two types. The discharges manifest as filamentary transfer of electrons (microdischarges) in a gap between two electrodes. To evenly distribute the microdischarges, a dielectric insulator must be used to separate the metallic electrodes and to prevent arcing.

7.4.4 Electrochemical Ozone Production

A significant effort has been dedicated to the investigation of electrochemical ozone production (EOP) [46–75]. In the EOP systems, water molecules are split into H_2, O_2, and O_3. In the most EOP methods, the hydrogen gas will be removed to leave oxygen and ozone as the only reaction products. Therefore, EOP can achieve higher dissolution in water without other competing gases found in the corona discharge method, such as nitrogen gases present in ambient air. The development of equipment that can efficiently produce highly concentrated ozone in aqueous media and has a long operation lifetime is important for industrial use. The advantages of the EOP method are simple system design, low-voltage operation, the potential for very high efficiency equating to very high concentrations in the gas and liquid phases, no need for gas feeds of any description and robust, proven cell and system technology through the established chlor-alkali industry. Ozone was produced first by electrolysis in 1840 [76]. Electrochemical ozone generation can be achieved under acidic conditions with high current efficiencies, i.e., 20–35%, but this has generally required low temperatures, e.g., 0 to –64 °C [76].

The development of EOP systems using a polymer electrolyte membrane (PEM) has shown significant progress, because such systems enable the realization of a compact apparatus and low-voltage operation. The ozone formation reactions in the direct water electrolysis systems are as follows:

Anode:

$$2H_2O \rightarrow O_2 + 4H^+ + 4e^-, \quad E^0 = 2.229V \qquad (7.15)$$

$$3H_2O \rightarrow O_3 + 6H^+ + 6e^-, \quad E^0 = 1.511V \qquad (7.16)$$

$$O_2 + H_2O \rightarrow O_3 + 2H^+ + 2e^-, \quad E^0 = 2.075V \qquad (7.17)$$

Cathode:

$$2H^+ + 2e^- \rightarrow H_2, \quad E^0 = 0.00V \tag{7.18}$$

The EOP systems can work in two main operation modes of batch and continuous (Figs. 7.3 and 7.4).

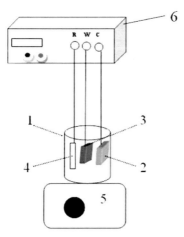

Figure 7.3 Batch setup: 1. electrochemical cell, 2. counter electrode, 3. working electrode, 4. reference electrode, 5. magnetic stirrer, 6. coulometer.

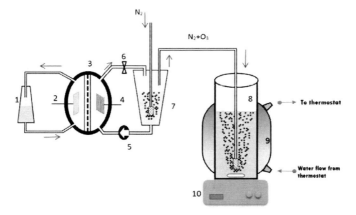

Figure 7.4 Continuous setup: 1. catholyte storage vessel; 2. cathode (Ti/PtO$_2$); 3. nafion 117 membrane; 4. anode (Ti/Sn-Sb-Ni); 5. rotary pump; 6. sampling port for ozone measurement; 7. anolyte storage vessel; 8. ozonation vessel; 9. water jacket; 10. magnet stirrer.

The performance of the EOP process is dependent on factors such as temperature, nature of the electrode material, composition of the supporting electrolyte, etc. [77–81]. Literature reports suggest the use of electrode materials having a high over potential for the oxygen evolution reaction (OER; e.g. b-PbO$_2$) combined with an SE containing fluoro-compounds and low temperature to obtain the best efficiency for EOP. However, fundamental aspects explaining the influence of fluoro-anions on the process are still subject of controversy [71, 79, 82]. The complexity of the phenomena governing EOP was recently demonstrated by the present authors [83], which showed that the morphological characteristics (roughness, porosity) of the electrode material significantly affect the current efficiency and require further investigation of the electrode preparation parameters to optimize EOP, thus reducing energy demands.

7.5 Anode Materials

One of the main interests and an open research subject in studies of EOP systems is working on the preparation of effective anodes. Up to now, many anode substances have been tested for ozone production in a PEM system. The first report of highly concentrated ozone water production using a PEM and a PbO$_2$ anode catalyst was published by Stucki et al. [84]. Since then, the effect of following anodes in the electrochemical ozone generation has been studied: Fe-doped PbO$_2$ catalyst electrodeposited on a Ti tube [85], IrO$_2$–Nb$_2$O$_5$ catalyst deposited on a Ti anode [86], Sb-doped SnO$_2$ anode catalyst [87], Ni- and Sb-codoped SnO$_2$ [88], TaO$_x$/Pt/Ti anode [89], B-doped diamond electrode (BDD) [90], and n-type TiO$_2$ film fabricated on a Pt/TiO$_x$/Si/Ti substrate [91, 92]. Some catalysts, such as PbO$_2$ and Sb, are harmful to the human body and cannot be used for systems where contact of the human body with ozone water is inevitable; the use of an electrolyte, such as HCl or H$_2$SO$_4$, is also undesirable for the same reason; diamond electrodes are inferior in forming water flow channels that enable good contact between ozone and water in the apparatus; and no catalyst lifetime longer than 500 h has been reported [93].

Thus, the inhibition of O$_2$ evolution is a first requirement for an efficient generation of O$_3$ at a reasonable current [94]. A suitable choice can be achieved of the electrode material or the use of additives that partially inhibit the oxygen evolution reaction (OER) via blocking of its active sites [83]. Physical and chemical stability of electrode are the second requirement since the generation of O_3 needs a very high anodic potential. Thus, materials for electrochemical generation of ozone must have a high overpotential for oxygen evolution and should also be stable to strong anodic polarization in the electrolyte [95]. Various anodes have been studied for electrochemical ozone production, such as platinum, diamond, alpha- and beta-PbO_2, Pd, Au, dimensionally stable anode (DSA), glassy carbon, SnO_2–Sb_2O_5, and Ni–Sb–SnO_2 [96–99]. Gold, DSA, and glassy carbon electrodes give low current efficiencies of less than 1% [88]. Platinum shows higher current efficiency from 6.5% to 35% but only at a very low temperature (−50 °C) and at the room temperature, the current efficiency fell to 0.5%. PbO_2 anodes can produce ozone at a current efficiency of 13% at room temperature, and for Ni–Sb–SnO_2 nanocomposites, a coat on titanium mesh shows higher current efficiency of ozone production (>36.5%) [100–102].

At present, a tin-based electrode with Sb and doped with other ions is a promising anode for the electrochemical treatment of wastewater because of its good conductivity, high catalytic activity, and high over potential for oxygen evocation, and low cost. However, the commercial application of a tin-based anode is hampered due to the deactivation of the electrode, which leads to a short lifetime of service. On these electrodes, one of the main secondary reactions is the evocation of the oxygen reaction during the process of the electrochemical oxidation of pollutants. The anodic oxygen-transfer reactions largely occur on the sites of the dopant in the semiconducting oxide mixture, which has a high overpotential of the evocated oxygen. Dopant ions can greatly improve the electrode's electrocatalytic activity and stability.

Tin dioxide is a semiconductor that the oxygen vacancies would donate electrons to its conduction band [103]. Also antimony (V) as donor (energy level ED) would donate electrons to the conduction band; thus, it could increase the conductivity by increasing the charge carrier density. Nickel (III), due to its lower valence than tin (IV), functioned as an acceptor (energy level EA) and would introduce positively charged holes in the valence band by accepting electrons from the bulk. The band diagrams for donor and acceptor were illustrated in Fig. 7.5.

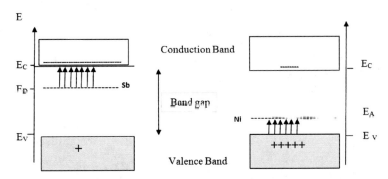

Figure 7.5 Band energy of diagram for semiconductor.

The electrons donated by antimony (V) and oxygen vacancies could be well compensated with the holes generated by nickel (III), and would be negatively charged around the nickel site. This negative charge would be trapped by adsorbed oxygen, that is, the oxygen would be strongly absorbed around the nickel site. The negative charge around nickel (III) would distribute more uniformly and decrease the oxygen absorption. While after donating out electrons, the antimony (V) sites would be positively charged and were always considered the Lewis acid sites, on which water may be oxidized to OH free radicals [104] and dioxygen may be formed quickly. In this process, on the antimony and nickel doped tin oxide electrode, due to the adjacent nickel site available, the dioxygen formed on antimony (V) will be transferred to and then adsorbed on the nickel site as an intermediate. The adsorbed dioxygen will react further with the hydroxyl free radical on adjacent antimony site to form HO_3 radical, which would be quickly oxidized to ozone by deprotonation. Figure 7.6 shows the mechanism for ozone generation on nanocomposite of N–Sb–SnO$_2$-coated Ti electrode.

Morphological characteristic was observed with scanning electron microscopy (SEM) and X-ray deffraction (XRD). The crystal size and structure of the composite are important factors for the efficiency of the anode, especially when the crystal size is decreased to the nanoscale. In Fig. 7.7, XRD spectra show that the average particle size of composite, L, was under 10 nm and it was confirmed in SEM images, see Fig. 7.8. The average grain sizes were calculated by Scherrer equation:

$$L = \frac{k \cdot \lambda}{\text{FWHM} \cdot \cos\theta} \tag{7.19}$$

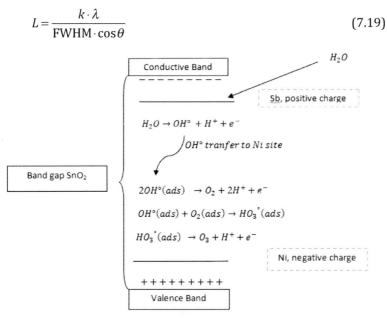

Figure 7.6 Mechanism for ozone generation on nanocomposite of N–Sb–SnO$_2$ coated on Ti electrode.

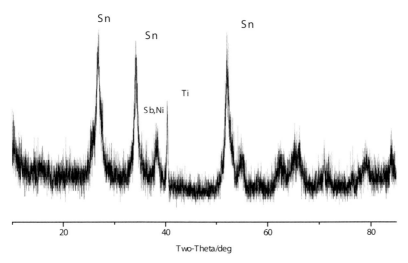

Figure 7.7 XRD spectra showing the average particle size of composite.

In Eq. (7.19), L is the average grain size (nm), k is Scherrer constant, λ is the wavelength of X-ray, FWHM is obtained from the diffraction peak width at half maximum, and θ is the diffraction angle (rad).

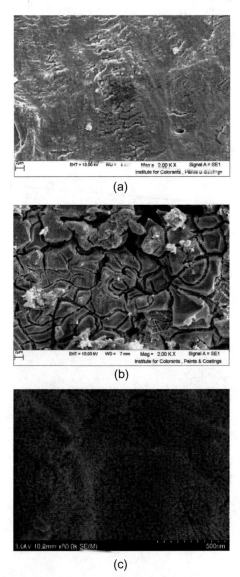

Figure 7.8 SEM (a) surface of net titanium without coating and (b, c) with coating of nanocomposite Sn–Sb–Ni.

The concentration of dissolved ozone was determined by a UV-Vis spectrophotometer; 1 mg L^{-1} dissolved ozone gave an absorbance of 0.098 at the wavelength of 258 nm. The UV spectrum was calibrated by standard indigo method [105]. Figure 7.9 shows UV spectrum of ozone that the gradual increase of the ozone UV peak. The first spectrum was collected after 1 min of electrolysis. The peak absorbance increases with electrolysis time because the ozone concentration increases during electrolysis.

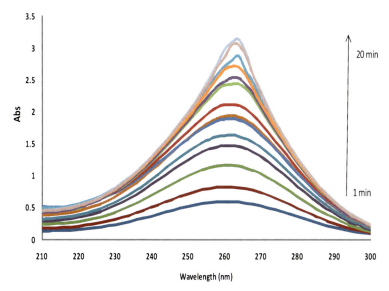

Figure 7.9 UV spectra solution with dissolved ozone concentration, applied voltage = 2.4 V vs. Ag/AgCl, C_{HClO4} = 1 M.

As shown in Fig. 7.10, a steady-state aqueous ozone concentration was reached after 20 min (1200 s) at 2.4 V vs. Ag/AgCl and different concentration of $HClO_4$. The dissolved ozone concentration depended on the applied voltage, concentration, and type of electrolyte. Result shows that the concentration of ozone varies in the range of 2 mg L^{-1} to above 27 mg L^{-1} at the experimental condition. These values of dissolved ozone are more than necessary concentration for most applications and are usually not attainable in the conventional CCD process [106]. Figure 7.10 is shown in aqueous ozone concentration.

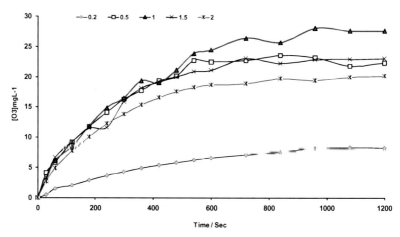

Figure 7.10 Ozone concentrations in water vs. electrolysis time at different concentration of HClO$_4$. O$_3$ was generated applying a constant potential of 2.4 V vs. Ag/AgCl.

7.6 Application of Electrochemically Generated Ozone

Application of ozone (O$_3$) as oxidant for degradation of dyestuff pollutants has been used by several researchers. It is evident that ozonation can achieve high color removal, reduce organic compounds, improve biodegradability, and ensure effective disinfection without byproduct residuals. Ozonation for decolorizing dyestuff wastewater has the following advantages: It (1) does not increase the volume of wastewater and sludge; (2) removes color and reduces the organic matter in one step; (3) needs little space and is easily installed on a site; and (4) is less harmful than other oxidative processes since no stock hydrogen peroxide or other chemicals are required on a site; and (5) residual ozone can be easily decomposed to oxygen. Reactive Blue 19 (RB-19) is a model anthraquinone reactive dye. A few past studies concerning ozonation of RB-19 solution were focused on degradation efficiency and color removal. Fanchiang and Tseng have reported the transformation of RB-19 under different ozonation conditions in 10 min of reaction time and found that partial oxidation was obtained, and also ozone gas was generated from pure oxygen by the ozone generator with classic corona method for degradation of RB-19 [107–109].

The degradation of RB-19 by in situ electrochemical generated ozone in acidic media at room temperature was investigated [109]. Nanocomposite coating times effect, electrolyte type, electrolyte concentration, voltage applied, and initial dye concentration have been optimized. The decolorization and degradation of the dyes via electrochemically generated ozone technology were investigated in several papers. Ozone was generated with high coulombic efficiency via electrochemical oxidation of water on anode electrode at different operating conditions. The electrode was characterized by CV and SEM. High concentration of generated ozone was obtained at 2.4 V vs. Ag/AgCl in 1 M $HClO_4$ after 10 min electrolysis time. Removal of color and COD were 100% and 48.2%, respectively (Fig. 7.11). Intermediate compounds were detected by integrated gas chromatography–mass spectrometry (GC/MS). Most intermediates have identified linear structures that are less harmful to the environment although some aromatic compounds have identified but their types are low.

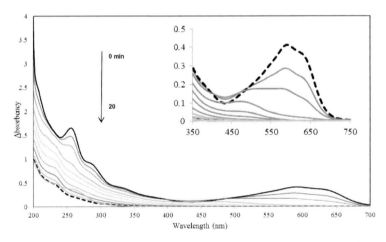

Figure 7.11 Change in the absorption UV/Vis spectra of RB-19 during ozonolysis; initial dye concentration is 50 mg L^{-1}.

References

1. Rice R. G., Netzer A. (eds). (1982). *Handbook of Ozone Technology and Applications*, Vol. 1, Ann Arbor Science, England.

2. Stucki S., Theis G., Kötz R., Devantay H., Christen H. (1985). *J. Electrochem. Soc.*, 132, 367.
3. Meng M. X., Hsieh J. S. (2000). *Tappi J.*, 83, 67.
4. Foller P. C., Tobias W. (1982). *J. Electrochem. Soc.*, 129, 506.
5. Foller P. C., Goodwin P. C. (1984). *Ozone Science Eng.*, 6, 29.
6. Ota K., Kaida H., Kamiya N. (1987). *Denki Kagaku Oyobi Kogyo Butsuri Kagaku*, 55, 465.
7. Shepelin V. A., Bubak A. A., Potapova G. F., Kasatkin E. V., Roginskaya Y. E. (1990). *Elektrokhimiya*, 26, 1142.
8. Wen T. C., Chang C. C. (1992). *J. Chin. Inst. Chem. Eng.*, 23, 397.
9. Tatapudi P., Pallav J. M. (1993). *J. Electrochem. Soc.*, 140, 3527.
10. Wen T. C., Chang C. C. (1993). *J. Electrochem. Soc.*, 140, 2764.
11. Feng J., Johnson D. C., Lowery S. N., Carey J. J. (1994). *J. Electrochem Soc.*, 141, 2708.
12. Potapova G. F., Kasatkin E. V., Nikitin V. P., Shestakova O. V., Blinov A. V., Mazanko A. F., Sorokin A. I., Asaturov S. A. (1995). *Bashk. Khim. Zh.*, 2, 65.
13. Zhou Y., Wu B., Gao R., Zhang H., Jiang W. (1996). *Yingyong Huaxue*, 13, 95.
14. Perret A., Haenni W., Niedermann P., Skinner N., Comminellis Ch., Gandi D. (1997). *Proc. Electrochem. Soc.*, 97.
15. Kim J. K., Choi B. S. (1997). *Hwahak Konghak*, 35, 218.
16. Chernik A. A., Drozdovich V. B., Zharskii I. M. (1997). *Russ. J. Electrochem.*, 33, 259.
17. Chernik A. A., Drozdovich V. B., Zharskii I. M. (1997). *Russ. J. Electrochem.*, 33, 264.
18. Fateev V. N., Akel'kina S. V., Velichenko S. V., Girenko D. V. (1998). *Russ. J. Electrochem.*, 34, 815.
19. Babak A. A., Amadelli R., Fateev V. N. (1998). *Russ. J. Electrochem.*, 34, 149.
20. Katsuki N., Takahashi E., Toyoda M., Kurosu T., Lida M., Wakita S., Nishiki Y., Shimamune T. (1998). *J. Electrochem. Soc.*, 145, 2358.
21. Meng M. X., Hsieh J. S. (2000). *Tappi J.*, 83, 67.
22. Foller P. C., Goodwin M. L. (1984). *Ozone Sci. Eng.*, 6, 29.
23. Stucki S., Theis G., Kötz R., Devantay H., Christen H. (1985). *J. Electrochem. Soc.*, 132, 367.

24. Stucki S., Baumann H., Christen H. J., Kötz R. (1987). *J. Appl. Electrochem.*, 17, 773.
25. Kowal A. L., Świderska-Broż M. 2007. *Oczyszczanie wody. Wydawnictwo Naukowe*, PWN, Warszawa.
26. Langlais B., Reckhow D. A., Brink D. R. 1991. *Ozone in Water Treatment: Application and Engineering*, Lewis Publishers, Chelsea, MI, USA.
27. Degremont. 2007. *Water Treatment Handbook*, Degremont, Rueil-Malmaison Cedex, France.
28. Letterman R. D (ed). 1999. *Water Quality and Treatment, A Handbook of Community Water Supplies*, fifth edition, American Water Works Association/McGraw-Hill, Inc.
29. Nawrocki J., Biłozor S (eds). 2000. *Uzdatnianie Wody. Procesy chemiczne i biologiczne*, PWN, Poznań-Warszawa.
30. Alsheyab M. A. T., Muñoz A. H. (2007). *Desalination*, 217, 1–7.
31. Hoigne J., Bader H. (1976). *Water Res.*, 10, 377–386.
32. Chiron S., Fernandez-Alba A., Rodriguez A., Garcia-Calvo E. (2000). *Water Res.*, 34, 366–377.
33. Sanchez L., Peral J., Domenech X. (1998). *Appl. Catal. B: Environ.*, 19, 59–65.
34. Langlais B., Reckhow D. A., Brink D. R (eds). 1991. *Ozone in Water Treatment: Application and Engineering*, Lewis Publishers, Chelsea, Michigan, USA.
35. Nawrocki J., Biłozor S. 2000. Uzdatnianie wody, Procesy chemiczne i biologiczne, PWN, Warszawa-Pozna´n,
36. Kowal A. L., Owiderska-Bróż M. 1997. Oczyszczanie wody, PWN, Warszawa-Wrocław.
37. Masschelein W. J. 1992. *Unit Processes in Drinking Water Treatment*, Marcel Dekker, New York.
38. Camel V., Bermond A. (1998). *Water Res.*, 32, 3208.
39. Huang C. P., Dong C., Tang Z. (1993). *Waste Manage.*, 13, 361.
40. Andreozzi R., Caprio V., Insola A., Marotta R. (1999). *Catal. Today*, 53, 51.
41. Hoigné J., Stucki S. (ed). 1988. *The Chemistry of Ozone in Water, Process Technologies for Water Treatment*, Plenum Press, New York.
42. Nawrocki J., Kasprzyk-Hordern B. (2010). *Applied Catalysis B: Environmental*, 99, 27–42.
43. Kim J. B., Yousef A. E., David S. (1999). *J. Food Protect.*, 62, 1071.

44. Han S. D., Jung D., Singh K., Chaudhary R. 2004. *Indian J. Chem. Sect A, Vol.,* 43, 1599.
45. Dohan, J. M., Masschelein W. J. (1987). *Ozone Sci. Eng.,* 9, 315–334.
46. Foller P. C., Goodwin M. L. (1984). *Ozone Sci. Eng.,* 6, 29.
47. Stucki S., Theis G., Kötz R., Devantay H., Christen H. (1985). *J. Electrochem. Soc.,* 132, 367.
48. Stucki S., Baumann H., Christen H. J., Kötz R. (1987). *J. Appl. Electrochem.,* 17, 773.
49. Babak A. A., Amadelli R., De Battisti A., Fateev V. N. (1994). *Electrochim. Acta,* 39, 1597.
50. Chernik A. A., Drozdovich V. B., Zharskii I. M. (1997). *Russ. J. Electrochem.,* 33, 259.
51. Beaufils Y., Bowen P., Comninellis C., Wenzed C. 1998. *Proc. Electrochem. Soc.,* 617, 730.
52. Beaufils Y., Comninellis C., Bowen P. 1999. *Electrochem. Eng. Symp. Series,* 145, 191.
53. Rice R. G., Netzer A. (eds). 1982. *Handbook of Ozone Technology and Applications,* vol. 1, Ann Arbor Science, UK.
54. Amadelli R., De Battisti A., Girenko D. V., Kovalyov S. V., Velichenko A. B. (2000). *Electrochim. Acta,* 46, 341.
55. Foller P. C., Tobias W. (1982). *J. Electrochem. Soc.,* 129, 506.
56. Ota K., Kaida H., Kamiya N. (1987). *Denki Kagaku Oyobi Kogyo Butsuri Kagaku,* 55, 465.
57. Shepelin V. A., Babak A. A., Potatova G. F., Kasatkin E. V., Roginskaya Y. E. (1990). *Élektrokhimiya,* 26, 1142.
58. Wen T. C., Chang C. C. (1992). *J. Chin. Inst. Chem. Eng.,* 23, 397.
59. Tatapudi P., Pallav J. M. (1993). *J. Electrochem. Soc.,* 140, 3527.
60. Wen T. C., Chang C. C., (1993). *J. Electrochem. Soc.,* 140, 2764.
61. Feng J., Johnson D. C., Lowery S. N., Carey J. J. (1994). *J. Electrochem. Soc.,* 141, 2708.
62. Potapova G. F., Kasatkin E. V., Nikitin V. P., Shestakova O. V., Blinov A. V., Mazanko A. F., Sorokin A. I., Asaturov S. A. (1995). *Bashk. Khim. Zh.,* 2, 65.
63. Zhou Y., Wu B., Gao R., Zhang H., Jiang W. (1996). *Yingyong Huaxue,* 13, 95.
64. Perret A., Haenni W., Niedermann P., Skinner N., Comninellis C., Gandi D. (1998). *Proc. Electrochem. Soc.,* 97.

65. Kim J. K., Choi B. S. (1997). *Hwahak Konghak*, 35, 218.
66. Chernik A. A., Drozdovich V. B., Zharskii I. M. (1997). *Russ. J. Electrochem.*, 33, 264.
67. Fateev V. N., Akel'kina S. V., Velichenko A. B., Girenko D. V. (1998). *Russ. J. Electrochem.*, 34, 815.
68. Babak A. A., Amadelli R., Fateev V. N. (1998). *Russ. J. Electrochem.*, 34, 149.
69. Katsuki N., Takahashi E., Toyoda M., Kurosu T., Lida M., Wakita S., Nishiki Y., Shimamune T. (1998). *J. Electrochem. Soc.* 145, 2358.
70. Kötz E. R., Stucki S. (1987). *J. Electroanal. Chem.*, 228, 407.
71. Velichenko A. B., Girenko D. V., Kovalyov S. V., Gnatenko A. N., Amadelli R., Danilov F. I. (1998). *J. Electroanal. Chem.*, 454, 203.
72. Amadelli R., Armelao L., Velichenko A. B., Nikolenko N. V., Girenko D. V., Kovalyov S. V., Danilov F. I. (1999). *Electrochim. Acta*, 45, 713.
73. Foller P. C., Tobias W. (1981). *J. Phys. Chem.*, 85, 3231.
74. Foller P. C., Kelsall G. H. (1993). *J. Appl. Electrochem.*, 23, 996.
75. Da Silva L. M., De Faria L. A., Boodts J. F. C. (2001). *Pure Appl. Chem.*, 73, 1871.
76. Foller P. C., Tobias C. W. (1982). *J. Electrochem. Soc.*, 129, 506.
77. Foller P. C., Goodwin M. L. (1984). *Ozone Sci. Eng.*, 6, 29.
78. Rice R. G., Netzer A. (eds). 1982. *Handbook of Ozone Technology and Applications*, vol. 1, Ann Arbor Science, UK.
79. Foller P. C., Tobias W. (1982). *J. Electrochem. Soc.*, 129, 506.
80. Kotz E. R., Stucki S. (1987). *J. Electroanal. Chem.*, 228, 407.
81. Foller P. C., Tobias W. (1981). *J. Phys. Chem.*, 85, 3231.
82. Velichenko A. B., Girenko D. V., Kovalyov S. V., Gnatenko A. N., Amadelli R., Danilov F. I. (1998). *J. Electroanal. Chem.*, 454, 203.
83. Da Silva L. M., De Faria L. A., Boodts J. F. C. (2001). *Pure Appl. Chem.*, 73, 1871.
84. Stucki S., Theis G., Kotz R., Devantay H., Christen H. J. (1985). *J. Electrochem. Soc.*, 132, 367.
85. Feng J., Johnson D. C., Lowery S. N., Carey J. J. (1994). *J. Electrochem. Soc.*, 141, 2708.
86. Santana M. H. P., De Faria L. A., Boodts F. C. (2004). *Electrochim. Acta*, 49, 1925.
87. Cheng S., Chan K. (2004). *Electrochem. Solid State Lett.*, 7, 134.

88. Wang Y., Cheng S., Chan K., Li X. Y. (2005). *J. Electrochem. Soc.*, 162, D197.
89. Awad M. I., Seta S., Kaneda K., Ikematsu M., Okajima T., Ohsaka T. (2006). *Electrochem. Commun.*, 8, 1263.
90. Arihara K., Terashima C., Fujishima A. (2006). *Electrochem. Solid State Lett.*, 9, D17.
91. Kitsuka K., Kaneda K., Ikematsu M., Iseki M., Mushiake K., Ohsaka T. (2010) *J. Electrochem. Soc.*, 157, F30–F34.
92. Marselli B., Garcia-Gomez J., Michaud P. A., Rodrigo M. A., Comninellis C. (2003). *J. Electrochem. Soc.*, 150, D79.
93. Naya K., Okada, F. (2012). *Electrochim. Acta*, 78, 495.
94. Da Silva L. M., De Faria L. A., Boodts J. F. C. (2003). *Electrochim. Acta*, 48, 699.
95. Wang Y. H., Cheng S., Chan K. Y., Li X. Y. (2005). *J. Electrochem. Soc.*, 15, 197.
96. Arihara K., Terashima C., Fujishima A. (2007). *J. Electrochem. Soc.*, 154, E71.
97. Tatapuip P., Fenton J. W. (1993). *J. Electrochem. Soc.*, 140, 3527.
98. Foller P. C., Tobias W. (1981). *J. Electrochem. Soc.*, 85, 3231.
99. Chen Q. Y., Shi D. D., Zhang Y. J., Wang Y. H. (2010). *Water Sci. Technol.*, 62, 2090.
100. Christensen P. A., Lin W. F., Christensen H., Imkum A., Jin J. M., Li G., Dyson C. M. (2009). *Ozone Sci. Eng.*, 31, 287.
101. Basiri Parsa J., Abbasi M. (2012). *J. Solid State Electrochem.*, 16, 1011.
102. Basiri Parsa J., Abbasi M., Cornell A. (2012). *J. Electrochem. Soc.*, 159, D265.
103. Jolivet J. P. (2000). *Metal Oxide Chemistry and Synthesis: From Solution to Oxide*, John Wiley & Sons, New York.
104. He D., Mho S. (2004). *J. Electroanal. Chem.*, 568, 19.
105. Cheng S. A., Chan K. Y. (2004). *Electrochem. Solid State Lett.*, 7, D4.
106. Fernando J. B. (2004). *Ozone Reaction Kinetics for Water and Wastewater Systems*, CRC Press, Boca Raton.
107. Fanchiang J. M., Tseng D. H. (2009). *Chemosphere*, 77, 214.
108. Santana M. H. P., Da Silva L. M., Freitas A. C., Boodts J. F. C., Fernandes K. C., De Faria L. A. (2009). *J. Hazard. Mater.*, 164, 10.
109. Basiri Parsa J., Abbasi M. (2012). *J. Appl. Electrochem.*, 42, 435.

Chapter 8

Core–Shell Nanocomposites for Detection of Heavy Metal Ions in Water

Sheenam Thatai, Parul Khurana, and Dinesh Kumar
Department of Chemistry, Banasthali University,
Banasthali 304022, Rajasthan, India
dschoudhary2002@yahoo.com

Water pollution by heavy metals has become one of the most severe environmental problems today. In recent years, various methods for heavy metal detection from water have been extensively studied. The present article brings out a series of information on different varieties of core–shell nanocomposites such as SiO_2@Au, SiO_2@Ag, Ag@SiO_2, Au@SiO_2, etc., for water purification. These nanocomposites provide high surface area and specific affinity for heavy metal adsorption from aqueous systems. To date, it has become a hot topic to develop new technologies to synthesize different nanocomposites, to evaluate the detection of heavy metals under varying experimental conditions, to reveal the underlying mechanism responsible for metal removal based on modern analytical techniques (SEM, TEM, SERS, etc.) or mathematical models, and to develop materials of better applicability for practical use. This chapter mainly focuses

---

Nanocomposites in Wastewater Treatment
Edited by Ajay Kumar Mishra
Copyright © 2015 Pan Stanford Publishing Pte. Ltd.
ISBN 978-981-4463-54-6 (Hardcover), 978-981-4463-55-3 (eBook)
www.panstanford.com

on nanocomposites preparation, their physicochemical properties, adsorption characteristics and mechanism, as well as their application in heavy metal detection. These materials exhibited strong aptitude to detect metal ions such as Pb(II), Cd(II), Zn(II), etc. Detection by different nanocomposites showed that concentration level can be successfully detectable down to the acceptable level of environmental standard.

8.1 Introduction

Nature has evolved many complex and new mechanisms for materials synthesis [1]. Living organisms produce materials with physical properties similar to those of analogous synthetic materials with almost same phase and compositions. Therefore, scientists have improved the properties of composite materials unachievable with traditional materials and investigated composites with lower and lower fillers size, leading to the development of microcomposites to nanocomposites.

A composite is defined as a combination of two or more materials with different physical and chemical properties and distinguishable interface. There are many advantages of composites over many metal compounds such as high toughness, high specific stiffness, high specific strength, gas barrier characteristics, flame retardance, corrosion resistance, low density, and thermal insulation [2]. In most composite materials, one phase is continuous, which is called the matrix, while the other phase is known as the dispersed phase. Nanocomposites refer to the composites in which one phase has nanoscale morphology such as nanoparticles, nanotubes, or lamellar nanostructure [3, 4]. It means these multiphase materials must have at least one of its constituent phase with dimension less than 100 nm. In nanocomposites, covalent bonds, ionic bonds, van der Waals forces, and hydrogen bonding could exist between the matrix and filler components [4]. Macroscopic properties of a composite material mixed with nanometric fillers, i.e., properties of nanocomposites are determined by various factors, such as its composition, the characteristics of each component, the geometry of the filler, the filler dispersion, the filler–filler and filler–matrix interactions, and in some cases with the modification of the characteristics of the matrix itself [5]. These parameters have a great influence on the final properties of the nanocomposite and are strongly interconnected. The improvement in properties can be obtained when there is good interaction and dispersion between the nanoparticles and the matrix. Adsorption of composites onto the surface of filler results in restriction of mobility and serves to suppress chain transfer reactions [6].

8.2 Classification of Nanocomposites

We are exposed to many different kinds of composites on a daily basis. We can classify them in many ways such as by composition, inclusion size, and morphology. On the basis of the nature of the matrices, composites can be classified into four major categories as shown in Fig. 8.1:

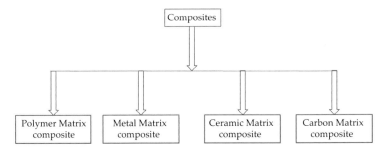

Figure 8.1 Classification of nanocomposites on basis of matrices.

Different types of nanofillers have been studied, including silica, clay, carbon black, alumina silicates, carbon nanotubes, and cellulose, which is of particular interest as it has rod-like shape. As nanocomposite properties depend on the good dispersion of the filler in the matrix, so it is necessary to develop technical methods to characterize the nanodispersion. They are also classified according to the fillers type and their inclusion with matrix as represented in Fig. 8.2.

Inorganic layered materials exist in wide range of variety, and they possess well-defined, ordered intralamellar space accessible by foreign species. This ability enables them to act as matrices for polymers and thus yielding interesting hybrid nanocomposite materials.

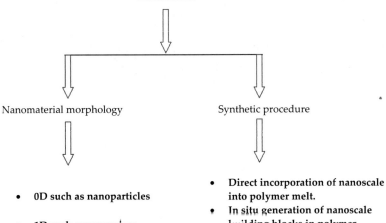

Figure 8.2 Classification of nanocomposites on basis of fillers.

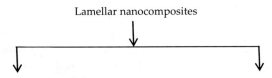

In intercalated nanocomposites, the polymer chain is in alternate with the inorganic layers in a fixed compositional ratio and has a well-defined number of polymer layers in the intralamellar space, whereas in exfoliated nanocomposites, the number of polymer chains between the layers is almost continuously variable and the layers stand more than 100 Å apart. The intercalated nanocomposites are also more compound-like because of the fixed polymer–layer ratio, and they are interesting for their electronic and charge transport properties, but exfoliated nanocomposites are more interesting as they have superior mechanical properties. Our work focuses on the lamellar class of intercalated organic/inorganic nanocomposites and namely those systems that exhibit electronic properties in at least one of the components. Surface-related properties, such as optical, magnetic, electronic, catalytic, mechanical, and chemical properties can be obtained by advanced nanostructured manufacturing, making them attractive for industrial applications in high-speed machining, optical applications [7], and magnetic storage devices because of their special mechanical, electronic, magnetic, and optical properties due to size effect.

8.3 Methods for Preparation of Nanomaterials as Nanofillers

Preparation of nanoparticles is an important branch of materials science and engineering. A variety of methods for the preparation of nanoparticles are available, which are typically grouped into two main categories: The "top-down" approaches involve the construction of nanoparticles from a much larger solid, while the "bottom-up" approaches involve the condensation of atoms or molecular components in a gas or liquid phase. Top-down approaches remove, reduce, or subdivide bulk material to make nanomaterials and all classified methods are presented in Fig. 8.3.

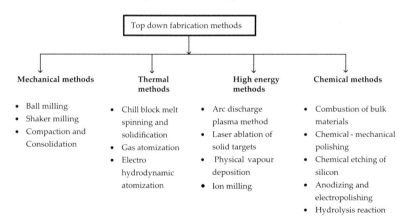

Figure 8.3 Classification of nanomaterials by "top-down approaches."

In bottom-up fabrication approaches, atoms or molecules combine to form nanomaterials. Therefore, these methods are considered to be additive. Bottom-up methods are divided according to the phase within which the process occurs and shown in Fig. 8.4.

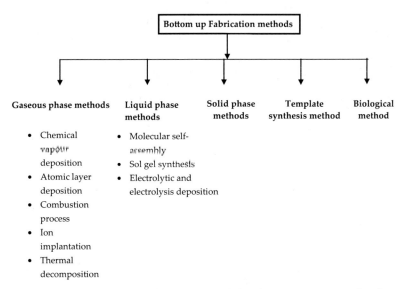

Figure 8.4 Classification of nanomaterials by "bottom-up approaches."

A variety of materials are prepared by the abovementioned methods behaving as nanofillers. Ceramic materials serve as the matrix host material in a composite. These types of materials are composed of both metallic and nonmetallic elements such as Al_2O_3, TiO_2, SiO_2, TiN, SiC, etc. They serve either as matrix host material in a composite or as the minority inclusions in metal or polymer matrix. These nanoscale ceramic materials are used as catalysts, photoconductors, and as template materials. Semiconductors, including CdS, CuS, and PbS, find their way into nanocomposites as they have optical, electronic, and sensing functions. Others include Au nanometal clusters embedded in Si matrix, which will generate a unique property as optical response based on surface plasmon. Different types of polymers such as polyesters, polyethene, nylon, PVC, are lightweight materials and are easily mixed to form new materials. Inorganic carbon materials such as graphite, fullerenes, carbon nanotubes, and diamondoids have great contribution to carbon-reinforced epoxy resin composites. Researchers are trying to incorporate carbon nanotubes into composites with all major classes of engineering materials as host. Biologically based materials such as carbohydrates, proteins, lipids, and nucleotides are also showing a great value as components in composites. Proteins are nontoxic and can be easily functionalized. Some nanofillers synthesized by authors are Fe_3O_4, TiO_2, CdS, PbS, CuS, and SiO_2 nanoparticles.

8.3.1 Fe_3O_4 Nanoparticles

Fe_3O_4 nanoparticles prepared by using $FeCl_3 \cdot 6H_2O$, $FeCl_2 \cdot 4H_2O$, and HCl (12 mol) were dissolved in deionized water, which is degassed with nitrogen gas before use. Then, NaOH solution was added drop wise under vigorous stirring at 80 °C. Obtained Fe_3O_4 nanoparticles were washed with deionized water 3–4 times and resuspended in deionized water. These magnetic nanoparticles were further used in preparation of Fe_3O_4@Au nanocomposites.

8.3.2 TiO_2 Nanoparticles

Titania particles were produced by the sol-gel method [8]. Tetraethyl orthotitanate, also known as titanium ethoxide (TEOT), was used as the titania precursor. In a typical synthesis procedure, ethanol and aqueous solution of metal salts such as NaCl or KBr were mixed and stirred for 15 min at room temperature in inert atmosphere. TEOT was added drop wise, and the mixture was continuously stirred for another 30 min. Turbidity appears instantly as soon as TEOT is added, and the mixture was allowed to age for 150 min. The dispersion of formed TiO_2 particles was then filtered and washed with ethanol.

8.3.3 CdS, PbS, and CuS Nanoparticles

With ethanol and thioglycerol (TG) used as organic solvent and surface capping agent, respectively, CdS, PbS, and CuS nanoparticles were prepared by the reaction between metals and elemental sulfur in ethanol at 70 °C. Then the prepared solution was heated at 70–80 °C for 25 min under stirring. The solution was centrifuged and the precipitate was washed with deionized water. CdS, PbS, and CuS powder were obtained after drying in vacuum oven for 8 h [9].

8.3.4 SiO_2 Nanoparticles

Silica particles were synthesized using Stöber procedure [10]. Hydrolysis and successive condensation of silica precursor tetraethylorthosilicate (TEOS) was carried out in alcoholic medium in a base catalyzed reaction using ammonium hydroxide. In a typical preparation method, a mixture of ethanol, distilled water, and ammonium hydroxide was stirred for 30 min to form a homogeneous solution. TEOS was added to this solution, and the total solution was stirred for 3 h to get white color precipitate. This precipitate was washed with water several times to remove traces of NH_4OH and dried to collect in powder form. This procedure yields highly monodispersed silica particles. Variation of TEOS/ethanol ratio and NH_4OH concentration produces particles of different sizes. Particle sizes ranging from 50 to 500 nm could be synthesized by altering the reactant concentration.

To improve the dispersion of the nanoparticles, physical or chemical methods are needed to alter the physical, chemical, and mechanical properties, and the surface structure of nanoparticles. The surface modification of nanoparticles can be divided into partial chemical modification, mechanical and chemical modification, external membrane modification, and high-energy surface modification. Surface modification in general is done through physical adsorption coating or grafting, depending on the properties of the particle surface. By the chemical reactions between the nanoparticle surface and the modifier, the surface structure and state of nanoparticles are changed. Surface chemical modification of nanoparticles plays a very important role to reduce the agglomeration. Due to modifier adsorption or bonding on the particle surface, which reduces the surface force of hydroxyl groups, the hydrogen bonds between particles are eliminated to prevent the formation of oxygen bridge bonds when nanoparticles are drying, thereby preventing the occurrence of agglomeration.

8.4 Methods for Preparation of Nanomaterials as Matrix

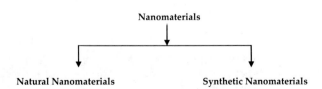

Nanomaterials occurring in natural environments without artificial modification are known as natural nanomaterials, and those prepared synthetically are known as artificial nanomaterials. A wide variety of natural materials are formed by living organisms by complex biochemical processes. Different types of minerals with complex nanostructure give rise to optical, absorbent, and catalytic properties. For example, clay, aluminosilicates, zeolites, etc., are complex minerals with 3D crystalline structure with network of interconnected tunnels and cages, since they have negatively charged framework with the pores that hold cations easily. Multicellular organisms can synthesize protein matrix to hold cell together in a structured body. These matrices provide support and strength to tissues of all living organisms. Paper is the most common and familiar substance made from cellulose residue of digested stems, wood, and fibers. Cellulose contained in it has good properties absorbency and strength. All these natural materials are built from bottom up and layer by layer incorporation of many different atoms and molecules into the structure, which give rise to their properties that they can behave as good adsorbents and catalysts.

Colloids, clusters, quantum dots, nanotubes, nanorods, nanowires, nanocylinders, nanospheres, aerogels, and polymers are the type of synthetic nanomaterials well prepared by bottom-up approaches and further can be used for preparation of nanocomposites. Gold metal has a great history as it can be converted to clusters, quantum dots, nanotubes, nanorods, nanowires, nanocylinders, and nanospheres. Synthesis of germanium nanowires had been accomplished at 1 atm and 400 °C [11]. Researchers have found straightforward methods to synthesize Fe-Pt nanowires with their length, diameter, and composition controlled [12]. Mobil crystalline materials (MCMs) are synthetic zeolites with pore diameter less than 2 nm and large up to 50 nm. These can be used as molecular sieves, adsorbents, and as shape-selective catalysts [13]. There are different polymers yet prepared, which can act as matrices. Inorganic polymers include strains of clay, sand, and fibers and many synthetic organic polymers, including adhesives, fibers, plastics, rubber, etc. Biological polymers include polysaccharides, proteins, cellulose, etc. Some nanomaterials behaving as matrices synthesized by authors are Au and Ag nanoparticles.

8.4.1 Au Nanoparticles

Citrate reduction method is first reported by Turkevich [14] and is popularly used to generate spherical gold nanoparticles. Simply put, gold salt, reducing agent, and citrate are stirred in water, and metal nanospheres are reduced.

$2HAuCl_4 + 3C_6H_8O_7$ (citric acid) $\rightarrow 2Au + 3C_5H_6O_5$ (3-ketoglutaric acid) $+ 8HCl + 3CO_2$

During the process, the temperature, the ratio of gold to citrate, and the order of addition of the reagents control the size distribution of gold nanospheres. The most popular one for a long time has been the use of sodium citrate reduction of $HAuCl_4$ in water. Adding freshly prepared trisodium citrate in boiled $HAuCl_4$ will produce deep wine-red color solution. It is then centrifuged with water and ethanol and redispersed in water.

8.4.2 Ag Nanoparticles

Silver nanoparticles were synthesized by a chemical reduction method using trisodium citrate [14] as well as sodium borohydride [15]. Since sodium borohydride is a very strong reducing agent, reaction takes place almost instantly and small particles are produced. Reduction using sodium borohydride is done at room temperature, while trisodium citrate needs higher temperature (around 80 °C) for reduction. Tri sodium citrate ($Na_3C_6H_5O_7$) was added to the silver nitrate solution, and the resulting mixture was refluxed at 80 °C for 30 min, and a pale yellow colored solution of silver nanoparticles was formed.

8.5 Methods for Preparation of Nanocomposites

There is a wide range of nanocomposites that are well studied with various applications. Carbon-based nanocomposites are exciting engineered nanomaterials that have stimulated a great deal of interest over the past years [16]. They have different forms such as fullerenes, single and multi-walled CNTs (SWNTs and MWNTs, respectively), carbon nanoparticles, and nanofibers. In addition to the exceptional mechanical properties, they also possess superior thermal and electrical properties.

Core–shell nanocomposites represent one of the most interesting areas of material science as they have unique and tailored properties for several applications. There are many different types of core–shell nanocomposites. In this class, nanoparticles can be metal, metal oxide, or silica core with a polymer of organic material or organic shell. Some of the particles in this category are SiO_2/PAPBA [poly(3-aminophenylboronic acid)] [17], Ag_2S/PVA (polyvinyl alcohol), CuS/PVA [18, 19], Ag_2S/PANI (polyaniline) [20], and TiO_2/cellulose. Another type of core–shell nanocomposites are those in which the core of particles consists of organic compounds and can be polymers of organic compounds. The shell is inorganic and is a metal or silica or silicone. Structures that fall under the category of polymer/metal are polyethylene/silver and polylactide/gold [21–23]. These are used to improve properties of other materials. There are nanocomposites that have a polymeric core and a polymeric shell and are dispersed in a matrix, which can be any material whose property is to be modified. One of the materials in this category is polymethylmethacrylate (PMMA) coated antimony trioxide compounded with polyvinylchloride (PVC)/antimony trioxide composites [24]. The interaction between PMMA and PVC along with antimony trioxide enhances toughness and strength of PVC. Silica is a good candidate to prepare core–shell nanocomposites, and silica nanoparticles emerge particularly as a suitable matrix due to their surface functionality, thus allowing bioconjugation to bioactivate molecules for objectives such as monitoring and marking. Metal-silica core–shell particles have been prepared by the use of a silane coupling agent to provide the metal surface with silanol anchor groups [25]. Silica shells not only enhance the colloidal and chemical stability but also control the distance between core particles within assemblies through shell thickness [26].

There are many nanocomposites discovered yet, which are helpful in detection of heavy metal ions in water. Carbon nanotube/nano-iron oxide composites demonstrate it could be utilized for the removal of chromium (III) in water [27]. Due to a large surface area, small, hollow, as well as layered structures, carbon nanotubes (CNTs) have already been investigated as promising adsorbents for various organic pollutants and metal ions, and its adsorption capacity can be easily modified by chemical treatments. As shown in Fig. 8.5, Core–shell CdTe/ZnO@SiO$_2$ nanocomposites prepared by the encapsulation of CdTe quantum dots (QDs) and ZnO nanorods (NRs) with a layer of mesoporous SiO$_2$ shell exhibit very interesting photoluminescent behaviors after interactions with heavy metal ions such as Hg^{2+}, Pb^{2+}, and Cu^{2+} [28].

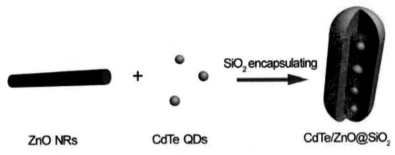

Figure 8.5 Schematic description of CdTe/ZnO@SiO$_2$ core–shell nanostructures [28].

Nanosized crystals of maghemite iron oxide (γ-Fe$_2$O$_3$) and further nanocomposite from γ-Fe$_2$O$_3$ with amberlite cationic exchange resin were synthesized. The nanocomposite shows remarkable adsorption efficiency in removal of some toxic metal ions such as Zn, Cu, and Cr [29]. Magnetic hydroxyapatite nanoparticle (MNHAP) adsorbents were synthesized and used for the removal of Cd^{2+} and Zn^{2+} from aqueous solutions [29].

Thermodecomposition process is developed to synthesize magnetic graphene nanocomposites decorated with core@double-shell crystalline nanoparticles, which are composed of crystalline iron core, iron oxide inner shell, and amorphous Si–S–O compound outer shell, which are found to possess unique capability to remove Cr(VI) very quickly and efficiently from wastewater [30]. Synthesis of alkanethiolate-functionalized core–shell Fe$_3$O$_4$/Au nanocomposites (Fe$_3$O$_4$@Au nanoparticles) that combine the advantages of core–shell magnetic nanoparticles with self-assembled monolayers (SAMs). Alkanethiolate has carboxylic acid (COOH) and methyl (CH$_3$) terminal groups, which can be easily self-assembled on Fe$_3$O$_4$@Au nanoparticles substrates [31]. Some core–shell nanocomposites synthesized by authors include SiO$_2$@Ag, SiO$_2$@Au, Fe$_3$O$_4$@Au, and Ag@Au.

8.5.1 SiO$_2$@Ag Core–Shell Nanocomposites

SiO$_2$@Ag core–shell nanocomposites were synthesized in a multistep process. Silica particles were synthesized by the Stöber method; on these particles, silver was coated in the second step. The method exploits the presence of opposite charges on the surface of silica particles and silver ions. Hence silver nanoparticles can be deposited on the surface of silica particles by electrostatic attraction. Silver nitrate was reduced in presence of silica particles using trisodium citrate (C$_6$H$_5$O$_7$Na$_3$). These particles as shown in Fig. 8.6 (a) were centrifuged and washed with water. Yellow color precipitate was again redispersed in water; the thickness can be enhanced by repeating the second step. Coating thickness can also be controlled by modifying the reaction conditions such as amount of silica particles added. Surface plasmon resonance of as formed metal nanoparticles as well as their core-shell composites are depicted in Fig. 8.6 (b).

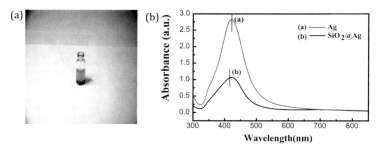

Figure 8.6 (a) Yellow colored colloidal solution and (b) surface plasmon resonance of Ag nanoparticles and SiO$_2$@Ag nanocomposites.

8.5.2 SiO$_2$@Au Core–Shell Nanocomposites

In order to prepare SiO$_2$@Au particles, first the silica particles were functionalized using 3-amino propyl triethoxy silane (APTES). APTES molecules have one OH end, and the other end has NH$_2$. Therefore, it can bond to silica through oxygen and gold via nitrogen atom. Functionalization of silica particles was performed by using APTES in C$_2$H$_5$OH·H$_2$O with silica particles to it. The resulting solution is vigorously stirred at 65 °C for 4 h. The solution was centrifuged, and the precipitate was washed with water. Gold solution, NaOH, and functionalized silica particles in water were stirred at 75 °C for 10 min. These form the gold seeds with silica particles. Finally, the above solution was centrifuged, washed with water, and redispersed in 40 mL water. Silica-gold seed solution and gold hydroxide solution were mixed and stirred with NaBH$_4$ [32]. Surface plasmon resonance of as formed gold nanoparticles and their nanocomposites with silica are given in Fig. 8.7.

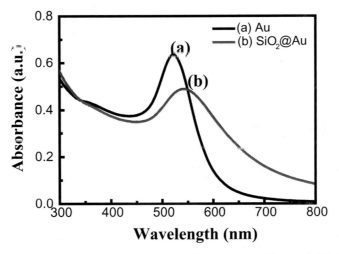

Figure 8.7 Surface plasmon resonance of Au nanoparticles and SiO$_2$@Au nanocomposites.

8.5.3 Fe$_3$O$_4$@Au Core–Shell Nanocomposites

The deposition–precipitation (DP) method can be used to synthesize oxide-supported gold particles. The coverage is controlled by several factors such as concentration of HAuCl$_4$, volume-to-mass ratio of HAuCl$_4$ to the support, base type (NaOH, NH$_4$OH, urea, TMAOH), reaction time and temperature, filtration, washing and drying steps, and calcinations temperature. Selection of the appropriate alkaline solution is especially critical for controlling the yield, dispersion, and size of the gold crystals. Fe$_3$O$_4$ nanoparticles with NH$_2$OH·HCl was diluted in TMAOH and at 80 °C, and 0.1% HAuCl$_4$ was added drop wise into the solution under vigorous stirring and nitrogen gas protection along with sodium citrate. The mixture was stirred for 3 h after the addition. During the whole process, temperature was maintained at 80 °C and nitrogen gas was used to prevent the intrusion of oxygen Optical images of as formed Fe$_3$O$_4$ and their core-shell nanocomposites with Au are shown in Fig. 8.8.

Characterization of Nanomaterials and Nanocomposites | 205

Figure 8.8 Colloidal solution of Fe$_3$O$_4$@Au and Fe$_3$O$_4$.

8.5.4 Ag@Au Core–Shell Nanocomposites

In boiled AgNO$_3$ aqueous solution, 1% trisodium citrate was added and diluted with ultrapure water and NH$_2$OH·HCl. Further, addition of HAuCl$_4$ drop wise (2 mL/min) and stirring vigorously lead to the formation of Ag@Au nanocomposites. It is stirred for 40 min, centrifuged with water then methanol, and redispersed in methanol SEM images of Ag@Au nanocomposites at different scales are shown in Fig. 8.9.

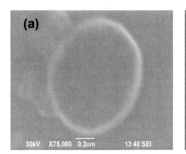
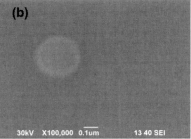

Figure 8.9 SEM images of Ag@Au nanocomposites.

8.6 Characterization of Nanomaterials and Nanocomposites

Nanomaterials and nanocomposites have to be characterized so as to ensure their formation, crystal structure, defect structure if any, structural phase, chemical composition, chemical bonding, etc. There are 700 single-signal characterization techniques and 100 multi-signal techniques [33]. Characterization of nanoparticles and core–shell nanocomposites was carried out using various techniques such as optical probe characterization techniques, electron probe characterization techniques, scanning probe characterization techniques, and spectroscopic characterization techniques. A brief description of these techniques is given in the following sections. Sample preparation, which is very important and varies for different techniques, is also explained briefly.

8.6.1 Optical Probe Characterization Techniques

Optics has played a central role, providing not only sufficient resolution to resolve morphological features at the cellular level, but also wavelength sensitivity, which enables the detection and localization of the optical signal from spectroscopically specific molecules or probes. Human eyes have a great depth of focus, but they are not efficient at viewing into nanostructured materials. Binoculars are useful for viewing morphology such as clumps. Optical microscopes are generally used for observing micron-level materials with reasonable resolution. Further, magnification cannot be achieved through optical microscopes due to limit in wavelength of light [34]. The probe in optical methods is visible light of wavelength within 400–800 nm range. Compound microscopes are also used, but the resolution is limited to few microns. Dynamic light scattering is an optical technique used to measure particle size.

8.6.2 Electron Probe Characterization Techniques

Electron probe characterization techniques use high-energy electron beams for imaging, chemical analysis, and determination of material structure. The mostly used electron probe methods are scanning electron microscopy (SEM) and transmission electron microscopy (TEM). Scanning electron microscopy (SEM) is an electron microscope that images the sample surface by scanning it with high-energy electrons. SEM creates magnified image using light waves [2]. SEM requires that the sample should be conductive for the electron beam to scan the surface and nonconductive samples are coated with layer of conductive material by low vacuum sputter coating or high vacuum evaporation. Nonconducting samples can also be analyzed using environmental SEM (ESEM) or field emission gun SEM (FE- SEM).

Transmission electron microscopy (TEM) is a technique where beam of electrons interacts and passes through the sample. An image is obtained as the electrons are transmitted from the sample and focused by objective lenses. The sample prepared for TEM must be in form of thin foil so that electron beam should penetrate from it. High-resolution TEM is another mode of TEM in which imaging of crystallographic structure of sample can be done at atomic level.

SEM and TEM instruments are coupled to other types of analytical tools known as EDX or EDS, i.e., energy dispersive X-ray analysis. EDX is used to analyze near-surface elements and their proportion at different positions, i.e., overall mapping of the sample.

8.6.3 Scanning Probe Characterization Technique

All SPM techniques have two things common: (1) a finely sharpened probe tip and (2) a system that enables the tip to scan over the surface of the sample. Scanning tunneling microscope (STM) is an instrument for forming the surface images. Here fine probe tip is scanned over the surface of conducting sample with piezoelectric crystal at distance of 0.5–1 nm. It is a powerful technique providing facilities for characterization and modification of variety of samples.

Atomic force microscopy (AFM) is another technique for measuring nanometer-scale surface roughness and for viewing surface nanotexture on different types of materials surface. AFM is a nondestructive technique with 3D spatial resolution. Here fine probe with a sharp tip near the end of beam is scanned across the surface of sample using piezoelectric scanners. It has three scan modes, namely, contact mode, noncontact mode, and tapping mode. AFM is used to explore nanostructures and properties of sample.

8.6.4 Spectroscopic Characterization Technique

There are many kinds of primary photon probe techniques in which UV spectroscopy based on absorption and emission of photons is the most common. Ultraviolet spectrometers consist of light source, reference and sample beams, monochromator, and a detector. Wavelength for UV-visible methods is within nanoscale. The UV spectrum for a compound is observed by exposing the sample to ultraviolet light from a light source. Parameters such as absorption, transmission, reflection, intensity, extinction, and wavelength are measured using UV-visible spectrometers.

Raman and Fourier transform infrared spectroscopy (FT-IR) are techniques that measure molecular and phonon vibrations of the sample. Symmetric vibrations are exclusive to Raman spectroscopy, and asymmetric vibrations reside in the domain of infrared analysis. The laser light interacts with phonons in the system and results in shifting of energy of laser phonons. This shift in energy gives information of the phonon modes of system. Raman Effect occurs only when light impinges on a molecule, interacts with electron cloud of the bonds of that molecule, and the incident photon excites one of the electrons to a virtual state. It provides the fingerprints by which sample can be identified in the range of 500–2000 cm^{-1}.

8.7 Sensing and Detection Using Smart Nanocomposites

Groundwater is water found beneath the surface of earth. Groundwater sources are cleaner than surface water sources. Mostly contaminants found in water are synthetic organic compounds, heavy metals, inorganic salts, biological pathogens, etc. The factors involved in drinking water quality were not well understood by ancients since there are some evidences of water treatment using charcoal, sunlight and through boiling in ancient years. Filtration was used to reduce turbidity and microbial contaminants such as those causing typhoid [35]. Modern techniques such as aeration, ion exchange, flocculation, and carbon filtration were designed to remove chemical species.

Recently, designing molecularly imprinted materials is one of the most promising approaches to explore sensing and detection applications. The formation of bulk metal from nanosized constituents improves many properties of metal. The control of material mechanical properties has been significant in the development of sensors. The area of sensor technology is very broad, and no simple classification is completely adequate. Detection of trace elements such as organic contaminants, explosive residues, and metal ions is an intellectually challenging task in science and engineering. It has great impact on society and the environment.

Sensors can be categorized by either their chemical composition or their principle of operation [36, 37]. Sensors are widely used in all aspects of everyday life, and their industrial applications include areas such as automotive, environment, and biomedicine [38]. The design and development of sensors requires many conditions such as the choice of the sensing element, issues of signal amplification, and measurement as well as suitable packaging of the sensor [39]. The sensing element refers to a material that fundamentally interacts with the target analyte and can also serve as a transducer by converting the interaction into a measurable output signal [40]. So "smart materials," which are able to respond to changes in their environment, can be used in chemical sensors wherein a chemical change in the environment is transduced into a signal [41]. One of the limitations in the use of "smart materials" is that the variations in environmental conditions and in analyte concentration need to be large to produce measurable responses.

For gold and silver nanoparticles, the surface plasmon absorption is one of the most attractive of these size-dependent properties because it leads to unique optical properties in the visible spectrum of light. As localized surface plasmon resonance (LSPR) of gold or silver nanoparticles exhibits red shifts upon forming nanoparticle aggregates because of interparticle plasmonic coupling, so the spectral shift of metal nanoparticles allows for the development of colorimetric sensors based on metal nanoparticles responsive to specific metal ions [42]. These colorimetric nanosensors can address intrinsic limitations of small molecular dye-based fluorescent sensors such as fluorescence quenching by metal ions and poor photostability. However, their practical uses exhibiting strong absorption of visible light are still problematic.

So recently, interest in composite materials consisting of metal nanoparticles and polymers in the form of networks has also increased. The addition of metal nanoparticles with unique properties to a polymer leads to reversible manipulation of these properties and greatly expands the range of material properties [39, 43]. The properties of the polymer- metal nanocomposites depend on the metal nanoparticle size, shape, and concentration as well as the interaction with the polymer matrix [43]. Metal nanoparticles are of great interest since they can show size-dependent properties different from those exhibited by the bulk metal. These effects mostly occur between 1 nm and 10 nm in diameter and their role in water treatment is shown in Fig. 8.10. Electronic, optical, catalytic, and thermodynamic properties have been observed to be the most significant size-dependent properties of metal nanoparticles [38].

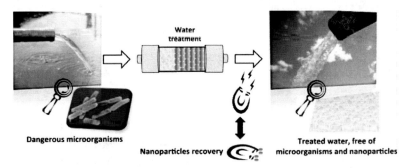

Figure 8.10 Use of environment-friendly nanocomposites for complex water treatment.

Presence of toxic ions such as lead, mercury, cadmium, zinc, etc., in drinking water, even in very small concentrations (parts per million concentration or higher), can cause serious health hazards. According to the United States Environment Protection Agency guidelines, the tolerance limit for lead, mercury, and cadmium in drinking water is no greater than 0.015 mg/L [26].

Earlier, gold nanoparticles have been used as sensor for colorimetric detection. Mirkin et al. have used two batches of gold particles that were attached to noncomplementary DNA oligonucleotides. When these particles were mixed with a solution containing a complementary DNA sequence, change in color (from red to blue) was observed because of the formation of self-assembled aggregates [44]. Similar approach was used for the detection of heavy metal ions [45], using gold nanoparticles capped with 11-mercatoundecanoic acid, which is a heavy metal ion receptor. These functionalized metal nanoparticles, which render red color, upon addition of salt solution turn to blue due to aggregation of particles. This color change or aggregation process is reversible via addition of a strong metal ion chelator such as EDTA. These responses could be observed spectroscopically using linear extinction as well as aggregation sensitive nonlinear Rayleigh scattering.

Detection and extraction of endosulfan, a commonly used pesticide, were done using gold and silver nanoparticles [46]. The interaction of endosulfan with gold and silver nanoparticles has been monitored by spectrophotometry, and it was found that gold nanoparticles are more sensitive toward endosulfan. Enhanced sensitivity of SPR band of SiO_2@Au nanocomposites compared to that of gold colloids has been reported [47]. Such enhanced sensitivities enable nanocomposites as an ideal candidate for detection of analytes present in drinking water.

SiO_2@Ag nanocomposites show enhanced SPR band than silver nanoparticles; therefore, a detection carried out using silver nanocomposites will be much more sensitive toward smaller quantity present in drinking water. With this idea, a test for the detection of heavy metal ions such as Pb^{2+}, Cd^{2+}, and Zn^{2+} using silver nanocomposites of size 220 nm (20 nm shell thickness) was developed. Solution of salt prepared in millipore water and its variable concentration was added to the nanoshell dispersion. SiO_2@Ag shows an intense SPR band at 443 nm. The optical changes that occur in the SPR band of SiO_2@Ag nanocomposites upon addition of these ions are monitored using a UV-Vis absorption spectrometer.

Silver nanocomposites synthesized by reduction method using trisodium citrate are stabilized by an electric double layer arising from adsorbed citrate and/or nitrate ions and corresponding cations. This layer is very sensitive toward electron transfer interactions as well as adsorption on the surface of silver nanoshells. This sensitivity has been exploited for the detection of toxic ions present in the solution. Addition of these ions causes changes in the position and shape of plasmon absorption band along with its damping due to chemisorption of these ions onto the surface of silver nanocomposites. Aggregation is observed between the particles, which is shown in Fig. 8.11(a & b).

Pb^{2+} ions get adsorbed on the surface of nanoshells, modifying the dielectric constant of the medium surrounding them, which results in the damping of plasmon absorption band. However, it is expected from Mie theory that a change in refractive index of the medium would lead to shift in the position of band maximum and would not influence the spectral width [48].

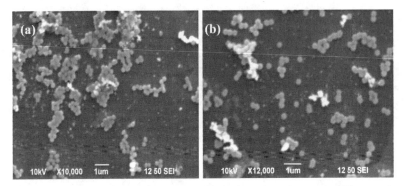

Figure 8.11 SEM images of aggregated particles in presence of Pb^{2+} ions.

Therefore, it is also possible that the adsorbed ions form a layer surrounding the nanocomposites and alter the optical properties of nanoshells themselves by changing their dielectric constant. Similar studies have been reported by Linnert et al. on damping of plasmon band by chemisorption of I$^-$, SH$^-$, and C$_6$H$_5$S$^-$ on the surface of silver nanoparticles. At higher concentrations (0.3 mL and 0.5 mL), plasmon band vanishes completely and also the color of dispersion changes from yellow to transparent.

Similar result could be observed for Zn^{2+} ions, in this case along with damping of plasmon band, gradual shifting of the band toward shorter wavelengths was also observed. However, in this case no change in the color of dispersion was observed. Such blue shift is observed when electrons were transferred from adsorbed ions to the metal particles. This phenomenon raises the density of free electrons in conduction band of the metal, hence increasing the plasma frequency of metal. Increase in plasma frequency eventually leads to blue shift in the extinction spectra (can be seen from Eq. 8.1)

$$\omega = \frac{4\pi e^2 N}{m}, \lambda = \frac{2\pi c}{\omega_p} \tag{8.1}$$

In a similar way, detection of Cd^{2+} ions in water is reported using silica-gold nanocomposites. Gold nanoparticles do not coalesce into each other to form a uniform smooth shell unlike some of the earlier reports [49–50]. However, they have "nanogaps" between them. Such gaps would be useful to provide more room for accommodation of ions as well as "hot spots" leading to better sensitivity. Hot spots can be attributed to enhanced electromagnetic field at the nanotips at such particles. Even the Au nanoparticles exhibit colorimetric detection of Cd^{2+} ions in water, but sensitivity of nanocomposites is ~50 times larger than that due to nanoparticles. Interaction between Cd^{2+} ions and gold nanoparticles or core–shell nanocomposites can be further investigated using Raman spectroscopy and results are depicted in Fig. 8.12. Raman spectroscopy is a very good technique to investigate the interactions at molecular level.

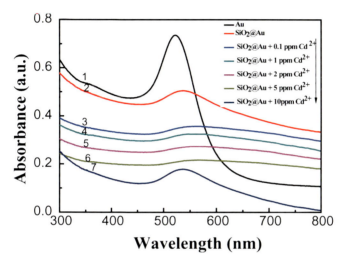

Figure 8.12 Changes in surface plasmon resonance peaks due to same addition of Cd^{2+} ions in silica–gold core–shell particle solution.

In Fig. 8.13, the Raman spectra of Au and SiO$_2$@Au are shown. The sensitivity of these surfaces, however, is quite low and failed to show any detectable signal of Au or SiO$_2$@Au covered ITO surfaces. Decrease in intensity for Au is noticeable when 5 ppm Cd^{2+} ions are present and for SiO$_2$@Au is 0.1 ppm.

Surface-enhanced Raman substrate consists of metallic micro- to nanosperical particulates on which analyte substrates are adsorbed. SERS is capable of 10^{12}–10^{14} enhancement of detection limits [51]. The wavelength of the excitation laser matches with the surface plasmon of nanocomposites. SERS is also able to detect many kinds of chemical agents as long as diagnostic peak exists for that material. Gold nanowire sensors able to detect ppb levels of Hg in water have been developed [52].

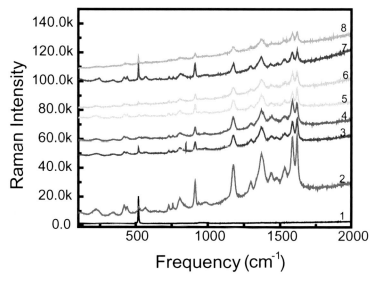

Figure 8.13 SERS spectra using CV (10^{-4}) as a molecule: (1) CV, (2) CV + Au, (3) CV + SiO$_2$@Au, (4) CV + SiO$_2$@Au + 0.1 ppm Cd^{2+}, (5) CV + SiO$_2$@Au + 1 ppm Cd^{2+}, (6) CV + SiO$_2$@Au + 2 ppm Cd^{2+}, (7) CV + SiO$_2$@Au + 5 ppm Cd^{2+}, (8) CV + SiO$_2$@Au + 10 ppm Cd^{2+}.

8.8 Conclusion

In this chapter, we provide an overview of information on different varieties of nanocomposites and with application of detection of metal ions in water. Nanocomposites have been a research field focused more from the last few years, and the advent of technologies provides a wide variety of novel methods for synthesizing particles at a micrometer or nanometer length scale. Plasmonic nanoparticles and their brilliant colors due to surface plasmon resonance absorption constitute a large ongoing research field. Composition, the characteristics of components used, the geometry of the filler, and the filler dispersion are some of the factors to be controlled. The improvement in properties of nanocomposites can be obtained when there is good interaction and dispersion between the nanoparticles and the matrix. Adsorption of composites onto the surface of filler results in restriction of mobility and suppresses chain transfer reactions with which these materials are able to detect foreign particles.

Acknowledgments

We are thankful to Professor Aditya Shastri, Vice Chancellor of Banasthali Vidyapith, for kindly extending the facilities of Banasthali Centre for Education and Research in Basic Sciences, sanctioned under CURIE programme of the Department of Science and Technology, New Delhi.

References

1. Xiaodong L., Chang W., Chao Y., Wang R., Chang M. (2004). Nanoscale structural and mechanical characterization of a natural nanocomposite material: The shell of red abalone, *Nanoletters*, 4, 613–617.
2. Horrocks A. R., Price D. (2008). *Advances in Fire Retardant Materials*, Woodhead Publishing Ltd, Cambridge.
3. Ratana D. (2005). *Epoxy Composites: Impact Resistance and Flame Retardancy: 16 (Rapra Review Report)*, Smithers Rapra Technology.
4. Friedrich K., Fakirov S., Zhang Z. (2005). *Polymer Composite—From Nano to Macro Scale*, Springer, New York.
5. Ljungberg N., Bonini C., Bortolussi F., Boisson C., Heux L., Cavaille J. Y. (2005). New nanocomposite materials reinforced with cellulose whiskers in atactic polypropylene: Effect of surface and dispersion characteristics, *Biomacromolecules*, 6, 2732–2739.
6. Kuljanin J., Marinović-Cincovic M., Stojanovic Z., Krkljes A., Abazovic D., Comor M. (2009). Thermal degradation kinetics of polystyrene/cadmium sulfide composites, *Polym. Degrad. Stab.*, 94, 891–897.
7. Ali S. M., Hasoon F. N., Hind A. F., Barghash A. H., Kazem S. H., Aljibori S. S. (2012). Technical overview for characterization and trends on the field of nanocomposite with their applications, *Eur. J. Sci. Res.*, 70, 159-168.
8. Eiden-Assmann S., Widoniak J., Maret G. (2004). Synthesis and characterization of porous and nonporous monodisperse colloidal TiO_2 particles, *Chem. Mater.*, 16, 6–11.
9. Wang X., Zhuang J., Peng Q., Yadong L. (2005). A general strategy for nanocrystal synthesis, *Nature*, 437, 121–124.
10. Stober W., Fink A., Bohn E. (1968). Controlled growth of monodisperse silica spheres in micron size range, *J. Colloid. Interf.*, 26, 62–69.

11. Gerung H., Boyle T. J., Tribby L. J., Bunge S. D., Brinker C. J., Han S. M. (2006). Solution synthesis of germanium nanowires using Ge^{2+} alkoxide precursor, *J. Am. Chem. Soc.*, 128, 5244–5250.
12. Wang C., Hou Y., Kim J., Sun S. (2007). A general strategy for synthesizing FePt nanowires and nanorods, *Angew. Chem. Int. Ed.*, 46, 6333–6335.
13. Kresge C. T., Leonowicz M. E., Roth W. J., Vartuli J. C., Beck J. S. (1992). Ordered mesoporous molecular sieves synthesized by a liquid crystal template mechanism, *Nature*, 359, 710–712.
14. Enüstün B. V., Turkevich J. (1963). Coagulation of colloidal gold, *J. Am. Chem. Soc.*, 85, 3317–3328.
15. Henglein A. (1989). Small-particle research: Physicochemical properties of extremely small colloidal metal and semiconductor particles, *Chem. Rev.*, 89, 1861–1873.
16. Brayner R. (2008). The toxicological impact of nanoparticles, *Nanotoday*, 3, 1–2.
17. Zhang Y. P., Lee S. H., Reddy K. R., Gopalan A. I., Lee K. P. (2006). Synthesis and characterization of core–shell SiO_2 nanoparticles/poly(3-aminophenylboronic acid) composites, *J. App. Polym. Sci.*, 104, 2743–2750.
18. Kumar R. V., Palchik O., Koltypin Y., Diamant Y., Gedanken A. (2002). Sonochemical synthesis and characterization of Ag_2S/PVA and CuS/PVA nanocomposite, *Ultrasonics Sonochem.*, 9, 65–70.
19. Francoise Q., Didier C., Francesco D., Corine G. (2006). Core–shell copper hydroxide-polysaccharide composites with hierarchical macroporosity, *Prog. Solid State Chem.*, 34, 161–169.
20. Jing S., Xing S., Yu L., Wu Y., Zhao C. (2007). Synthesis and characterization of Ag/polyaniline core–shell nanocomposites based on silver nanoparticles colloid, *Mater. Lett.*, 61, 2794–2797.
21. Naderi N., Sharifi-Sanjani N., Khayyat-Naderi B., Faridi-Majidi R. (2006). Preparation of organic–inorganic nanocomposites with core–shell structure by inorganic powders, *J. App. Polym. Sci.*, 99, 2943–2950.
22. Liu Y., Lee H., Yang S. (2005). Strategy for the syntheses of isolated fine silver nanoparticles and polypyrrole/silver nanocomposites on gold substrates, *Electrochimica Acta*, 51, 3441–3445.
23. Zhang Z., Wang F., Chen F., Shi G. (2006). Preparation of polythiophenes coated gold nanoparticles, *Mater. Lett.*, 60, 1039–1042.
24. Xie X., Li Y., Liu X., Mai W. (2004). Structure-property relationships of in situ PMMA modified nanosized antimony trioxide filled poly (vinyl chloride) nanocomposites, *Polymer*, 45, 2793–2802.

25. Marzán L., Giersig M., Mulvaney P. (1996). Synthesis of nanosized gold-silica core–shell particles, *Langmuir*, 12, 4329–4335.
26. Gupta V. K., Agarwal S., Saleh T. (2011). Chromium removal by combining the magnetic properties of iron oxide with adsorption properties of carbon nanotubes, *Water Res.*, 30, 1–6.
27. Yingying S., Xuebo C., Yang G., Peng C., Qingrui Z., S. Guozhen. (2009). Fabrication of mesoporous CdTe/ZnO@SiO$_2$ core/shell nanostructures with tunable dual emission and ultrasensitive fluorescence response to metal ions, *Chem. Mater.*, 21, 68–77.
28. Andra P., Avram N. (2012). Adsorption of Zn, Cu and Cd from waste water by means of maghemite nanoparticles, *UPB Sci. Bull.*, 74, 255–265.
29. Yuan F., Gonga J., Guang-Ming Z., Qiu-Ya N., Hui-Ying Z., Cheng-Gang N., Jiu-Hua D., Ming Y. (2010). Adsorption of Cd(II) and Zn(II) from aqueous solutions using magnetic hydroxyapatite nanoparticles as adsorbents, *Chem. Eng. J.*, 162, 487–494.
30. Jiahua Z., Suying W., Hongbo G., Sowjanya R., Qiang W., Zhiping L., Neel H., David Y., Zhanhu G. (2012). One-pot synthesis of magnetic graphene nanocomposites decorated with core@double-shell nanoparticles for fast chromium removal, *Environ. Sci. Technol.*, 46, 977–985.
31. Xiaoli Z., Yaqi C., Thanh W., Yali S., Guibin J. (2008). Preparation of alkanethiolate-functionalized core/shell Fe$_3$O$_4$@Au nanoparticles and its interaction with several typical target molecules, *Anal. Chem.*, 80, 9091–9096.
32. Khurana P., Thatai S., Wang P., Lihitkar P., Zhang L., Fang Y., Kulkarni S. K. (2012). Speckled SiO$_2$@Au core–shell particles as surface enhanced Raman scattering probes, *Plasmonics*, 8, 185–191.
33. Kelsall R. W., Hamley I. W., Geoghegan M. (2005). *Nanoscale Science and Technology*, John Wiley & Sons, London.
34. Joshi M., Bhattacharya A., Ali S. W. (2008). Characterization techniques for nanotechnology applications in textiles, *Indian J. Fibre Text. Res.*, 33, 304–317.
35. Environmental Protection Agency. (2000). *The History of Drinking Water*, Office of Water, Report EPA-816-F-00-006.
36. Davis J. J., Beer P. D. (2004). Nanoparticles: Generation, surface functionalization, and ion sensing, *Dekker Encyclopedia of Nanoscience and Nanotechnology*, 6, 2477–2491.

37. Gerlach G., Guenther M., Sorber J., Suchaneck G., Arndt K., Richter A. (2005). Chemical and pH sensor based on the swelling behavior of hydrogels, *Sens. Act. B*, 111–112, 555–561.
38. Klabunde K. J. (2001). *Nanoscale Materials in Chemistry*, John Wiley & Sons, Inc., New York.
39. Committee on New Sensor Technologies: Materials and Applications, National Materials Advisory Board, Commission on Engineering and Technical Systems, National Research Council. (1995). *Expanding the Vision of Sensor Materials*, National Academies Press, Washington, DC.
40. Galaev Y. (1995). 'Smart' polymers in biotechnology and medicine, *Russ. Chem. Rev.*, 64, 471–489.
41. Mingdi Y., Olof R. (2005). *Molecular Imprinted Materials: Science and Technology*, Marcel Dekker, New York.
42. Yin J., Wu T., Song J., Zhang Q., Liu S., Xu R., Duan H. (2011). SERS-Active nanoparticles for sensitive and selective detection of cadmium ion (Cd^{2+}), *Chem. Mater.*, 23, 4756–4764.
43. Bronstein L. M., (2004). Polymer colloids and their metallation, in *Dekker Encyclopedia of Nanoscience and Nanotechnology* (Schwarz J. A., Contescu C. I., eds), pp 2903–2915.
44. Marzán L., Giersig M., Mulvaney P. (1996). Synthesis of nanosized gold-silica core–shell particles, *Langmuir*, 12, 4329–4335.
45. Marzan L., Mulvaney P. (2003). The assembly of coated nanocrystals, *J. Phys. Chem. B*, 107, 7312–7326.
46. Sreekumaran Nair A, Tom R. T., Pradeep T. (2003). Detection and extraction of endosulfan by metal nanoparticles, *J. Environ. Monit.*, 5, 363–365.
47. Sun Y., Xia Y. (2002). Increased sensitivity of surface plasmon resonance of gold nanoshells compared to that of gold solid colloids in response to environmental changes, *Anal. Chem.*, 74, 5297–5305.
48. Kalele S. A., Tiwari N. R., Gosavi S. W., Kulkarni S. K. (2007). Plasmon assisted photonics at the nanoscale, *J. Nanophoton.*, 1, 012501–012505.
49. Pastoriza-Santos I., Gomez D., Perez-Juste J., Liz-Marzan M., Mulvaney P. (2004). Optical properties of metal nanoparticle coated silica spheres: A simple effective medium approach, *Phys. Chem. Chem. Phys.*, 6, 5056–5060.
50. Pol V. G., Gedanken A., Calderon-Moreno J. (2003). Gold–shell synthesis, deposition of gold nanoparticles on silica spheres: A sonochemical approach, *Chem. Mater.*, 15, 1111–1118.

51. Spencer K. M., Sylvia J. M., Janni J. A., Klein J. D. (2003). Advances in landmine detection using surface enhanced Raman spectroscopy, *SPIE Proceedings*, 3710
52. Keebaugh S., Nam W. J., Fonash S. J., Nanoscience and Technology Institute. (2007). Manufacturable, highly responsive nanowire mercury sensors, in *Symposium: Micro/nano-Technology for National Security Applications*, 20–24.

Chapter 9

Conducting Polymer Nanocomposite–Based Membrane for Removal of *Escherichia coli* and Total Coliforms from Wastewater

Hema Bhandari,[a] Swati Varshney,[a,b,c] Amodh Kant Saxena,[a] Vinod Kumar Jain,[b] and Sundeep Kumar Dhawan[b]
[a]*Polymeric and Soft Materials Section,*
CSIR—National Physical Laboratory, New Delhi 110012, India
[b]*Amity Institute of Advanced Research and Studies,*
Amity University, Noida 201303, India
[c]*Delhi Institute of Tool Engineering, Okhla Industrial Area, Phase II,*
Maa Anandmayee Marg, New Delhi 110020, India
skdhawan@mail.nplindia.org

Escherichia coli and total coliform are biological indices for pollution and contamination in water. Presence of fecal coliform bacteria in aquatic environment results from contamination of water with feces of humans or animals. At the time this occurred, the source water might have been contaminated by pathogens or disease-producing bacteria, which can also exist in fecal materials. These waterborne pathogenic bacteria remain a major health concern and are causing large number of deaths and hospitalization each year. Diarrhea caused by contaminated water is one of the serious problems. The fecal coliform bacteria may occur in ambient water as a result of the overflow of domestic sewage or nonpoint sources of human and other animal waste. This chapter describes the removal of *E. coli* and total coliform using conducting polymer nanocomposite-based membrane. Silver nanoparticles were prepared by the addition of silver nitrate and ethylene glycol solution in PVP/sodium hypophosphite. This membrane has been developed by in situ polymerization of pyrrole with silver nanoparticles into the activated carbon (AC) coated fiber in different concentration. Conducting polymer composite-based membranes were characterized by TGA analysis, FTIR spectroscopy, and XRD analysis, which indicate the inclusion of polymer and silver nanoparticles in the carbon fiber membrane. Transmission electron spectroscopy (TEM) and scanning electron microscopy (SEM) revealed the surface morphology as well as size of nanoparticle incorporated in the membrane matrix. Evaluation of the membrane for antimicrobial activity (removal of total coliform and *E. coli*) has been carried out by membrane filtration method. M-endo broth is used for selectively isolating coliform bacteria from water using membrane filtration method. Data were compared with uncoated AC fiber and only silver nanoparticles coated carbon fiber. Results revealed that conducting polymer-silver nanoparticle composite-based membrane showed markedly enhanced antimicrobial properties. This chapter will also describe the compatibility between the organic conducting polymer and inorganic silver nanoparticles in the carbon fiber matrix. Moreover, the synergistic role of conducting polymer and silver nanoparticles for antimicrobial activity for wastewater treatment will also be described.

9.1 Introduction

Water free from hazardous chemicals and pathogens is referred to as clean water, which is essential to human health. Unsafe water from all sources contributes significantly to the global burden of diseases, mainly through the waterborne transmission of gastrointestinal infections such as cholera, typhoid, hepatitis, and a wide range of agents that cause diarrhea. Waterborne diseases remain the leading cause of casualty in many developing countries. Pathogens such as *Escherichia coli* and other coliform bacteria are biological indices for pollution and contamination in water. Presence of fecal coliform bacteria in aquatic environment results from contamination of water with the feces of humans and animals. The most basic test for bacterial contamination of a water supply is the test for total coliform bacteria. Total coliform count gives a general indication of the hygienic condition of water supply. Total coliforms include bacteria found in water and soil. A large number of different bacteria, including, *E. coli, Enterobacter, Klebsiella, Citrobacter, Proteus*, and *Serratia* belong to the total coliform group. *E. coli* is considered to be the species of coliform bacteria and is the best indicator of fecal pollution and possible presence of pathogens.

Although use of disinfectants in drinking water can effectively control the pathogenic bacteria, but research in the past few decades exposed a difficulty between effective disinfection and formation of harmful disinfection by-products. Chemical disinfectants generally used by water industry, such as chloramines, chlorine, and ozone, can react with different components in natural water to produce disinfection by-products, many of which are highly carcinogenic. Krasner et al. reported a number of disinfection by-product [1]. High dose of chemical disinfectant for resistance of some pathogens such as *Crptosporidium* and *Giardia* leads to the formation aggravated disinfection by-products. Therefore, there is an urgent need to re-evaluate conventional disinfection methods and to consider innovative approaches that enhance effective antimicrobial activity and avoid the formation of disinfection by-products.

A variety of approaches for water purification are being pursued. The role of nanomaterials in the purification of water contaminated by toxic chemicals and pathogens is found to be an efficient method and extensively reported by many researchers [2–10]. Nanomaterials are brilliant adsorbents, sensor, and catalysts because of their large specific surface area and high reactivity. Recently, a lot of natural and engineering nanomaterials have been reported for wastewater treatment. Srivastava et al. [2] reported that carbon nanotubes as water filter can effectively remove bacterial pathogens, i.e., *E. coli* from contaminated water. Kang et al. [3] also reported the antimicrobial activity of carbon nanotubes. Number of nanomaterials have been reported for antimicrobial applications, such as chitosan [4], TiO_2 [5,

6], fullerene nanoparticles [7], ZnO nanoparticles [8, 9], and silver nanoparticles [10–12]. Unlike conventional antimicrobial materials, these nanomaterials are not strong oxidant and are inert in water; therefore, they are not expected to produce harmful disinfection by-product. But there are some limitations of these nanomaterials in water treatment technology, i.e., aggregation of these particles in water cancel out the benefit of nanomaterials. Due to their extremely small size, they may escape from the treatment system and enter the product water. Except silver nanoparticles (Ag NPs), most of them were found to be toxic under certain conditions [13].

Silver is found to be the most toxic element to microorganisms and are used in various applications such as wound dressing material, water disinfection, antimicrobial filters, chemical and gas filtration, air filtration, and protective cloth. Different polymers have been used to incorporate Ag-NPs for the production of antimicrobial membranes. Uses of impregnating nanoparticles into filter packing materials such as activated carbon (AC) or incorporation of these nanomaterials with another antimicrobial matrix is likely to overcome the limitation of nanomaterials. Use of AC as a matrix also plays an important role for holding a wide range of harmful contaminants from wastewater [14–19].

However, most research suggests that silver-impregnated carbon filters have very short-lived effectiveness in preventing bacterial growth in the water filter system [14–16]. Zodrow et al. [20] reported the effective use of polysulfonate ultrafiltration membranes impregnated with Ag-NPs for removal of *E. coli* and *Pseudomonas mendocina* bacteria strains. The antimicrobial device is based on the polymer nanocomposite containing metal nanoparticles, composed of polymer and metal nanoparticles. Silver nanoparticles have been widely used for water treatment due to its strong inhibitory and bactericidal effects as well as a broad spectrum of antimicrobial activities [21, 22]. Silver ions work against bacteria in a number of ways; it can interact with the thiol groups of enzyme and proteins that are important for the bacterial respiration and the transport of important substance across the cell membrane and within the cell [23, 24]. Silver nanoparticles comprise a high specific surface area and a high fraction of surface atoms that lead to high antimicrobial activity compared to bulk silver metal [23]. Polymer nanocomposites have also been attracted for antimicrobial applications due to their unique optical, electrical, and catalytic properties [25, 26] for biomedical devices [27, 28].

Conjugated polymers such as polypyrrole (PPY) and polyaniline (PANI) are generally employed in textile field for their electrical properties [29]. They can be easily produced by chemical oxidative polymerization in aqueous solutions of the monomer. Materials (e.g., fibers, fabrics) plunged in the polymerization bath are coated with an even and uniform layer of conjugated polymer by in situ chemical oxidative polymerization. Antimicrobial activity of conjugated polymers was first reported by Seshadri and Bhat [30, 31]. They deposited polyaniline on cotton fabrics by in situ chemical oxidative polymerization at low temperature. Recently, Zhang et al. [32] demonstrated the application of graphene-polypyrrole nanocomposite for the removal of perchlorate contamination from wastewater. Varesano et al. [33] reported the antimicrobial activity of polypyrrole-coated cotton fabrics. Chin-San Wu [34, 35] reported the antimicrobial activity of nanocomposites based on polyester/multiwalled carbon nanotubes and polyester/polyaniline composites. Hence, these materials were found to be highly effective for antimicrobial applications.

The present chapter describes the removal of total coliform and *E. coli* using conducting polymer (polypyrrole-Ag-NPs) nanocomposite-based membrane. The antimicrobial activity of embedded Ag-NPs in polymer matrices will be discussed; this membrane has been developed by in situ polymerization of pyrrole and Ag-NPs onto AC-coated fibers in different concentration. Activated carbon fiber is a material having high capacities of adsorption and catalysis [36, 37]. Silver nanoparticles were synthesized using silver nitrate and ethylene glycol solution in PVP/sodium hypophosphite/carbon fiber solution. Conducting polymer composite–based membranes were characterized by FTIR spectroscopy, TGA analysis, conductivity measurement, XRD analysis, transmission electron spectroscopy (TEM), and scanning electron microscopy (SEM). Evaluation of the membrane for antimicrobial activity (removal of total coliform and *E. coli*) has been carried out by membrane filtration method. M-endo broth has been used for selectively isolating coliform bacteria from water using membrane filtration method. Data will be compared with uncoated AC fiber and only Ag-NPs coated AC fiber. This chapter will also describe the compatibility between the organic conducting polymer and inorganic Ag-NPs in the carbon fiber matrix. Moreover, the role of conducting polymer and silver nanoparticles for inhibition of microbial activity in water will also be described.

9.2 Development of Polypyrrole-Silver Nanocomposites Impregnated AC Membrane

This membrane has been developed by in situ chemical oxidative polymerization of pyrrole and Ag-NPs onto the AC-coated fiber in different concentration.

9.2.1 Synthesis of Ag-NPs

Silver nanoparticles were prepared by using ethylene glycol as a reducing agent and polyvinyl pyrrolidone (PVP) as stabilizer. Ag-NPs were synthesized by dissolving 110 g PVP and 10 g sodium hypophosphite into 400 ml ethylene glycol. The solution was vigorously stirred at room temperature under ambient atmosphere and heated to 90 °C. To this solution, 1.0 M solution of silver nitrate in ethylene glycol was added with constant stirring. Formation of silver nanoparticles indicated by appearance of brown-colored suspension. The reaction was quenched, and suspension was rapidly cooled by adding chilled distilled water. Ag-NPs were separated by centrifugation and excess of PVP was removed using acetone as a solvent. Resulting Ag-NPs were dried under vacuumed at 50 °C. TEM micrographs revealed that Ag-NPs were spherical with diameter in the range of 5–10 nm.

9.2.2 Development of PPY Ag-NPs Impregnated AC Membrane

The AC-coated fibers were allowed to soak in the solution of pyrrole, sodium salt of p-toluene sulfonate (Na-PTS), and Ag-NPs (0.5 wt.% and 1.0 wt.%) for 1–2 h, which followed by the drop wise addition of oxidant (i.e., ferric chloride) with constant stirring for about 3–4 h. After deposition of PPY-Ag nanocomposites on AC-coated fibers, it was washed with water. PPY-Ag nanocomposites were embedded in

AC-coated membranes and were evaluated for antimicrobial activity of wastewater using membrane filtration method. The pictorial representation of chemical reaction is shown in Fig. 9.1, and the resulting conducting polymer nanocomposite was characterized by various techniques.

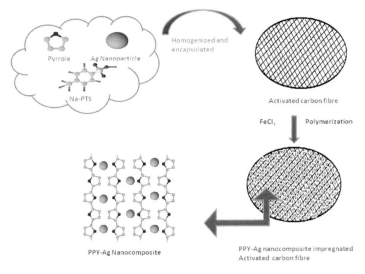

Figure 9.1 Pictorial representation for the development of PPY-Ag nanocomposite impregnated AC membrane.

9.3 Antimicrobial Activity Test Methods

Antimicrobial properties of PPY-Ag nanocomposite–based membrane were tested against *E. coli* and total coliform strains using membrane filtration method (method 8074). The membrane filtration method is a fully accepted and approved procedure for monitoring drinking water microbial quality in many countries. It is a fast way to estimate microbial growth in water. It is especially useful when evaluating large sample volumes or performing many coliform tests daily. This method consists of filtering a water sample on a sterile membrane filter made up of polytetrafluroethylene (PTFE) with a 0.45 μm pore size, which retains bacteria, incubating this filter on a selective medium, and enumerating typical colonies on the filter. Many media and incubation conditions for the membrane filtration method have been tested for optimal recovery of coliforms from water samples [38]. It is a quantitative procedure that uses membrane filters with pore sizes sufficient to retain the target organism. M-endo broth was used for selectively isolating coliform bacteria from wastewater. It is prepared according to the formula of Fifield and Schaufus [39]. It is recommended by the American Public Health Association (APHA) in standard total coliform membrane filtration method for analysis of water, wastewater, and food materials [40].

9.3.1 Membrane Filtration Method

In the initial step, an appropriate (i.e., 100 ml) wastewater sample volume containing coliforms was allowed to pass through a conducting polymer Ag nanocomposites impregnated AC fibers with pore size of about 0.1 mm. With this pore size, no coliform can be retained on its surface. Now the filtered water was passed through a conventional membrane (PTFE membrane) with pore size 0.45 μm, which is enough to retain the bacteria present. The filter was placed on an absorbent pad (in a Petri dish) saturated with a culture medium (m-endo broth), which is the selective media for coliform growth. The Petri dish containing the filter and pad was incubated for 24 h at the ambient temperature (35 °C). After incubation, the colonies that had grown were identified. Colonies grow in specific color and were manually counted. *E. coli* and total coliform bacteria form pink and blue colonies with a metallic sheen on m-endo broth medium containing lactose. Each experiment was repeated three times and counted total number of *E. coli* and total coliform using colony counter in terms of colony forming units (CFU) under magnifying glass and express as CFU/100 ml. The pictorial representation of methodology used in membrane filtration method is shown in Fig. 9.2.

In order to evaluate the efficiency of nanocomposite/AC-coated membrane, a large volume (about 25–30 L) of wastewater samples was passed through the single AC fiber, Ag, PPY, and PPY-Ag nanocomposite impregnated AC fiber and the total number of *E. coli* and total coliform was enumerated in terms of CFU/100 ml using PTFE membrane filter taking 100 ml of filtered water from bulk volume at each experiment. Each experiment was repeated three times.

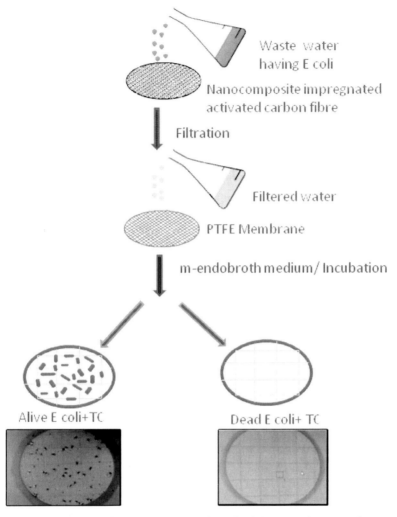

Figure 9.2 Pictorial representation of methodology used in membrane filtration method.

E. coli and total coliform density on a PTFE membrane filter was calculated as follows:

$$E.coli \text{ and } TC/100 \text{ ml} = \frac{\text{Coliform colonies counted}}{\text{Volume of sample filtered (ml)}} \times 100 \quad (9.1)$$

Moreover, reduction percentage ($R\%$) of coliform bacteria from coated and uncoated AC fibers can also be calculated using Eq. 9.2.

$$R(\%) = \frac{B-C}{C} \times 100, \qquad (9.2)$$

where B is the CFU/100 ml in membrane filter without treatment of wastewater and C is CFU/100 ml in membrane filter with treated water (using AC-coated or Ag-NPs or PPY or PPY-Ag nanocomposite impregnated AC fiber). The risk classification for *E. coli*/total coliform in household drinking water can be determined on the basis of total number of CFU/100 ml where 0 colony forming unit (CFU/100 ml) is "in conformity with WHO guidelines," 1–10 is "low risk," 10–100 is "intermediate risk," and 100–1000 is "very high risk." Nil CFU/100 ml is considered for drinking water.

9.4 Characterization of PPY-Ag Nanocomposite

9.4.1 Structural Characterization

9.4.1.1 FTIR spectra

Figure 9.3 shows the FTIR spectra of Ag-NPs, PPY, and PPY-Ag nanocomposites synthesized by in situ chemical polymerization. FTIR spectra of PPY showed the characteristic vibration bands at around 1321 cm^{-1} and 1512 cm^{-1}, are due to C–C or C=C stretching of quinoid structure and ring stretching of pyrrole ring, respectively. The peak at 1450 cm^{-1} is due to N–H (stretching) [41, 42]. Aromatic stretching vibration bands at 3060 cm^{-1} and 3025 cm^{-1} were observed in PPY. FTIR spectra of Ag-NPs revealed strong absorption peaks at 1380 cm^{-1}, 1151 cm^{-1}, 1600 cm^{-1}, and 1630 cm^{-1}, representing the presence of –NO$_2$, which may be from AgNO$_3$ solution and metal precursor involved in the synthesis of Ag-NPs. In the case of FTIR spectra of PPY-Ag nanocomposites, the typical absorption bands due to incorporation of Ag-NPs in PPY matrix were clearly observed (Fig. 9.3b), which showed the FTIR peak of Ag-NPs as well as PPY, indicating that the silver nanoparticles have been successfully encapsulated within the polypyrrole matrix during polymerization.

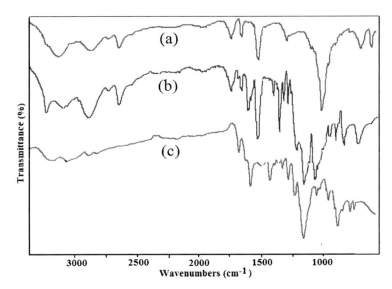

Figure 9.3 FTIR spectra of (a) Ag-NPs, (b) PPY-Ag nanocomposites, and (c) polypyrrole.

9.4.1.2 Conductivity measurement

Room temperature conductivity of PPY was found to be 0.16 S/cm. On incorporation of Ag-NPs (0.5 wt.% loading of Ag-NPs) in the PPY matrix, the room temperature conductivity value was found to increase from 0.16 S/cm to 1.45 S/cm indicating that the conductivity of PPY-Ag nanocomposites was better than that of conducting polymer (PPY). Incorporation of Ag-NPs enhances the conductivity and allows the PPY-Ag nanocomposite to more easily form an interconnected conductive pathway throughout the material. The higher conductivity value of about 1.75 S/cm has been observed in the 1.0 wt.% loading of Ag-NPs in PPY matrix, indicating that a higher amount of Ag-NPs has physically dispersed throughout the polymer matrix. This excess may lead to aggregate formation between the organic and inorganic phases.

9.4.1.3 X-ray diffraction analysis

Figure 9.4 shows the X-ray diffractogram of Ag-NPs, PPY, and PPY-Ag nanocomposites. The powder X-ray diffraction (XRD) pattern of PPY-Ag nanocomposites has further confirmed that Ag-NPs were incorporated into the PPP matrix, as evident by FTIR spectra and conductivity measurement.

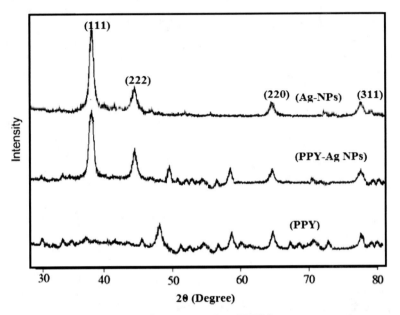

Figure 9.4 XRD pattern of Ag-NPs, PPY, and PPY-Ag nanocomposites.

The weak reflection of PPY centered at a 2θ value of approximately 28.2°, 49.1°, 65.1°, and 78.2° was characteristic of amorphous PPY [43, 44]. The sharp diffraction peaks at 2θ values of 38.5°, 43.0°, 65.5°, and 79.2° corresponded to Bragg's reflections from the (111), (222), (220), (311), and (222) planes of Ag-NPs. XRD pattern of PPY-Ag nanocomposites indicates that Ag-NPs have been successfully impregnated in the PPY matrix. These peaks indicate the crystalline nature of composites due inclusion of Ag-NPs. Nanocomposites containing higher amount of Ag-NPs (i.e., 1.0 wt.% loading) were found to be more crystalline, as evidenced by the intensity and sharpening of the peaks in XRD analysis.

9.4.2 Thermogravimetric Analysis

Figure 9.5 shows the thermogravimetric analysis of Ag-NPs, PPY, and PPY-Ag nanocomposites. The TGA curve of pure Ag-NPs only shows a small weight loss, occurring below 300 °C and can be ascribed to the elimination of water and ethanol.

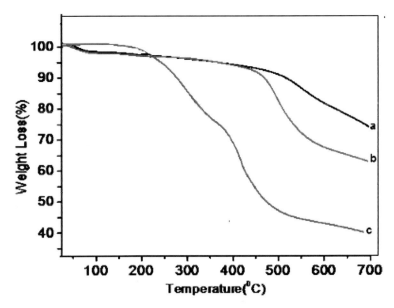

Figure 9.5 TGA curves of (a) Ag-NPs, (b) PPY-Ag nanocomposites, and (c) PPY at heating rate 10 °C/min in nitrogen atmosphere.

The weight loss of 14.5% observed until 350 °C in the TGA curve of Ag-NPs is attributed to the elimination of the adsorbed water and other solvents (Fig. 9.5a). TGA traces of PPY showed two weight loss steps. First a minor weight loss occurred up to 110 °C, which may be attributed to the loss of adsorbed water molecules. The second weight loss observed from 180 °C onward is attributed to degradation of polymer backbone. From the comparison of the TGA traces of PPY and PPY-Ag nanocomposites, it is observed that the first stage of weight loss was identical, whereas variation was observed in the second weight loss depending on the loading of Ag-NPs on PPY matrix during polymerization. It was observed that the PPY was thermally stable up to 180 °C, whereas PPY-Ag nanocomposites were thermally stable up to 285 °C. It indicates that the thermal stability of the composite has enhanced due to incorporation of Ag-NPs.

9.4.3 Antistatic Study

Static decay measurements were also performed on John Chubb Instrument (JCI 155 v5) charge decay test unit by measuring the time on applying the positive as well as negative high corona voltage of 5000 V on the surface of material to be tested and recorded the decay time at 10% cut-off [45]. A fast response electrostatic field meter observes the voltage received on the surface of sample and measurements were to observe how quickly the voltage falls as the charge is dissipated from the fiber. Graphs obtained from these experiments have been shown in Fig. 9.6, which show the decay of surface voltage and decay time. The surface voltage and surface charge received by the materials depend on the nature of materials [46, 47]. When positive or negative high corona voltage (i.e., 5000 V) was applied to the surface of the material, only a limited amount of voltage was received by the material depending on the nature of materials. When high corona voltage was applied on the surface of insulating material, only some voltage was drained away and greater amount of voltage was retained on its surface. This surface voltage decays at a particular time, and the decay time was measured at 10% cut-off. Hence the charge-retention capability of PPY-Ag nanocomposite–based AC fiber was lesser; thus, they quickly dissipate this surface charge as compared to PPY and Ag-NPs impregnated AC fibers. Due to the insulating nature of AC fibers, a lot of charges were found to be retained on its surface, and it was not able to dissipate the static charge up to 10% cut-off. Static decay time for Ag-NPs and PPY-impregnated AC fibers was found to be greater than 2.0 s at 10% cut-off, while PPY-Ag nanocomposites impregnated AC fibers showed the dissipation of static charge on around 0.6 s at 10% cut-off as shown in Fig. 9.6b. Therefore, better electrostatic charge dissipation (ESD) performance of PPY-Ag impregnated AC is due to the synergistic effect of AG-NPs and PPY. Moreover, better ESD performance of PPY-Ag nanocomposites indicates the proper encapsulation of Ag-NPs into the PPY matrix during in situ polymerization. Any material that showed a static decay time less than 2.0 s passes the criteria for its use as antistatic material. Based on the above observations, we can say that PPY-Ag nanocomposite impregnated AC fibers can also be used as an effective antistatic material.

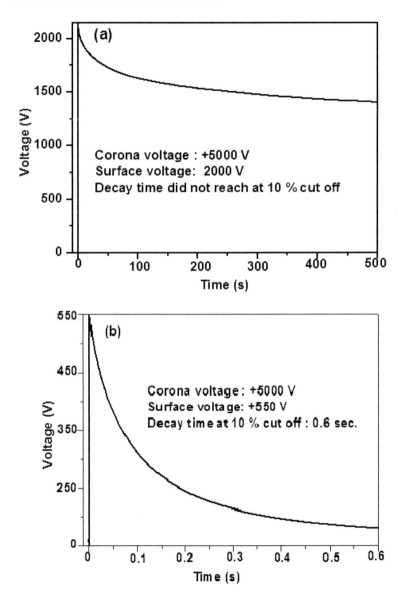

Figure 9.6 JCI graph of (a) AC fiber and (b) PPY-Ag nanocomposites/AC fiber.

9.4.4 Morphological Characterization

Figure 9.7 shows a typical SEM image of the PPY, Ag-NPs, and PPY-Ag nanocomposites. SEM image of Ag-NPs revealed the spherical morphology. Morphology of PPY was found to be a sponge-like structure. On incorporation of Ag-NPs in the PPY matrix, the morphology of PPY-Ag nanocomposites was modified as spherical in shape as shown in Fig. 9.7, which reveals the complete encapsulation of Ag-NPs into the polymer matrix.

Figure 9.7 SEM images of PPY, Ag-NPs, and PPY-Ag nanocomposites on AC fibers.

TEM images of these samples have been shown in Fig. 9.8a. The average dimension of Ag-NPs was found to in the range of 5–10 nm. The dimension of PPY-Ag nanocomposites was found to be in the range of approximately 20–50 nm due the incorporation of Ag-NPs with PPY matrix.

High resolution transmission electron microscopy (HRTEM) of the same samples has been shown in Fig. 9.8b–d, which reveals that the Ag-NPs are visible as dark spots inside the polypyrrole matrix. The HRTEM image (Fig. 9.8c,d) also reveals the microstructure and crystallinity of Ag-NPs and PPY-Ag nanocomposites, respectively.

The lattice fringes of spacing 0.232 nm corresponding to (111) plane confirms the presence of crystalline Ag-NPs in the amorphous polymer matrix [48].

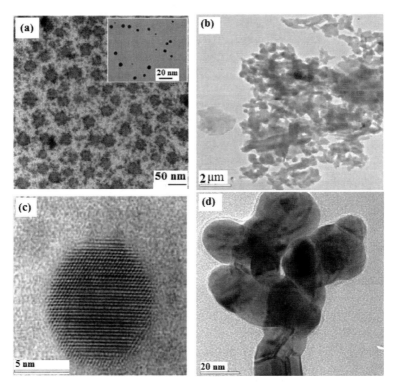

Figure 9.8 (a) TEM image of PPY-Ag nanocomposite, inset image indicates the TEM image of Ag-NPs, and HRTEM image of (b) PPY, (c) Ag-NPs, (d) PPY-Ag nanocomposites.

9.5 Antimicrobial Activity

Antimicrobial action of PPY-Ag nanocomposite impregnated AC fibers was checked by using bulk solution (~50 L) of *E. coli* and total coliform bacteria. Findings are summarized in Table 9.1. Initially, 100 ml of wastewater from bulk solution was passed through the membrane filter; the total number of CFU/100 ml in m-endo broth medium was found to be about 10^3 CFU/100 ml in untreated water (Fig. 9.9a).

Table 9.1 Antimicrobial efficiency of samples indicating total count of *E. coli* and total coliform (CFU/100 ml) and their reduction percentage

Sample name	Count of *E. coli* + TC (CFU/100 ml) approx value	Reduction (%) of *E. coli* + TC
Blank	>10³	Nil
AC coated fiber	800–900	20–10
Ag-NPs-impregnated AC	35	96
PPY-impregnated AC fiber	50	95
PPY-Ag (0.5 wt.% Ag) nanocomposite impregnated AC fiber	<10	99
PPY-Ag (1.0 wt.% Ag) nanocomposite impregnated AC fiber	Nil	100

Water treated with AC, on passing through the membrane filter, showed about 800–900 CFU/100 ml (Fig. 9.97b), while Ag-NPs and PPY-impregnated AC filtered water showed only 35 CFU/100 ml and 50 CFU/100 ml on membrane filter, respectively (Fig. 9.9c,d). Remarkable result was observed after passing the wastewater through PPY-Ag nanocomposites impregnated AC. On passing the wastewater (input water) through PPY-Ag nanocomposites impregnated AC fiber containing 0.5 wt.% loading of Ag-NPs, the output count of coliform bacteria was less than 10 CFU/100 ml. The output count was found to be zero on membrane filter after passing the wastewater through PPY-Ag (1.0 wt.% Ag) nanocomposite impregnated AC fiber, and no *E. coli* and total coliforms were detected in the treated water (Fig. 9.9f). Whereas the initial water sample (input water) showed overgrowth in m-endo broth on passing through uncoated fiber.

The percentage reduction of *E. coli* and total coliform on treated water has also been calculated using Eq. 9.2. Activated carbon coated fiber showed only 10–20% reduction in *E. coli* and total coliform from wastewater. While PPY and Ag-NPs impregnated AC showed 95–96% reduction in bacteria from wastewater. PPY-Ag (0.5wt.% loading of Ag-NPs) nanocomposites treated water, on passing through the PTFE membrane, revealed about 99% reduction in *E. coli* and total coliform. As shown in Table 9.1, 100% reduction in bacteria was observed using PPY-Ag (1.0 wt.% Ag) nanocomposite impregnated AC fibers.

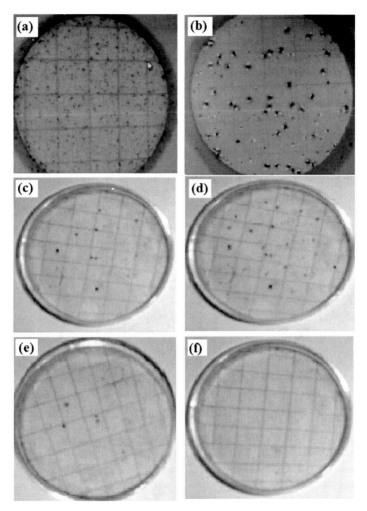

Figure 9.9 Pictures of membrane (PTFE) filter with and without treatment with containing *E. coli* and total coliform (CFU/100 ml) in m-endo broth medium (a) represent about 10^3 CFU/100 ml in PTFE membrane without treatment, (b) water treated with AC fiber, (c) water treated with Ag-NPs/AC fiber, (d) water treated with PPY/AC fiber, (e) water treated with PPY-Ag nanocomposites/AC fiber (0.5 wt.% loading of Ag NPs), and (f) water treated with PPY-Ag nanocomposites/AC fiber (1.0 wt.% loading of Ag-NPs).

Long-lasting efficiency of PPY-Ag nanocomposites coated AC has also been evaluated by passing 100 ml to 25 L wastewater through the fiber and calculating the total number of colonies after passing 100 ml of treated water from the membrane filter. Findings are summarized in Table 9.2, and graphical representations of the results have been shown in Fig. 9.10. According to IS 10500/2004, water containing up to 10 CFU/100 ml can be considered safe water.

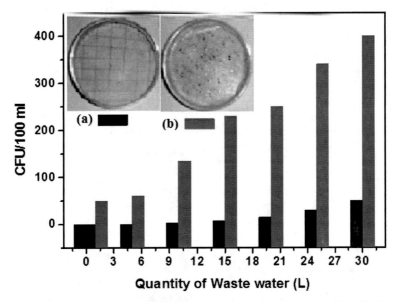

Figure 9.10 Effect of (a) PPY-Ag nanocomposite/AC fiber and (b) PPY/AC fiber on total count of *E. coli* and total coliform; inset figures indicate PTFE membrane showing CFU/100 ml when 15 L water is treated with (a) PPY-Ag nanocomposite/AC fiber and (b) PPY/AC fiber.

Results showed that PPY-coated AC was not found to be more efficient for wastewater treatment as compared to PPY-Ag-based nanocomposite coated AC fiber. Up to 20 L wastewater can effectively be treated using PPY-Ag nanocomposite coated AC fiber at a time. Although efficiency of PPY-coated AC fiber was only 1 L, after this limit, water was found to be unsafe and contaminated. This study revealed that nanocomposites containing Ag-NPs and PPY on AC matrix were stable and properly impregnated during polymerization and were not washed away by water flow, possibly due to its interaction with the nitrogen atom of the PPY ring. Formation of PPY-Ag impregnated AC-based membrane fiber having sufficient content of Ag-NPs results in long-term antibacterial effect.

Table 9.2 Effect of PPY-Ag nanocomposite/AC fiber on total count of *E. coli* and total count

Sample name	Quantity of water from bulk solution of wastewater	Count of *E. coli* + TC (CFU/100 ml) approx. value
PPY-impregnated AC fiber	1 L	50
	5 L	>50
	10 L	>100
	15 L	>200
PPY-Ag (1.0 wt.% Ag) nanocomposite impregnated AC fiber	1 L	Nil
	5 L	Nil
	10 L	1–2
	15 L	2–5
	20 L	>10
	25 L	>20

9.5.1 Antimicrobial Mechanism of PPY-Ag Nanocomposite Impregnated AC Fiber

It is presumed that PPY-Ag nanocomposites attach to the surface of the cell membrane disturbing permeability and respiration functions of the cell leading to uncontrolled transport through the plasma membrane and, finally, cell death. A number of studies have reported that Ag-NPs themselves act as an effective bactericidal material [49–51]. Antimicrobial studies indicated that only 93% and 95–96% reduction in *E. coli* and total coliform was observed using PPY-impregnated AC and Ag-NPs impregnated AC fiber. Remarkable results (i.e., 100% reduction in *E. coli* and total coliform) have been observed in case of PPY-Ag/ nanocomposite impregnated AC fiber.

Smaller Ag-NPs on PPY matrix having a large surface area available for interaction would give more antimicrobial effect. It is also presumed that PPY-Ag nanocomposites not only interact with the surface membrane of coliform, but can also penetrate inside the bacteria and kill them effectively. Research indicated that PPY can physically puncture bacteria and partially restrain the bacterial growth. In contrast, PPY-Ag nanocomposites have excellent bactericidal activity against *E. coli* and total coliform. PPY-Ag nanocomposites contained quaternary ammonium ion due to the PPY ring. The positive charges on the PPY ring seem to be responsible for the antimicrobial activity of conducting polymers.

SEM, HRTEM, and XRD studies indicate that during the polymerization of PPY-Ag nanocomposites, Ag-NPs were found to be embedded into the PPY matrix. In addition to the antimicrobial mechanism of PPY-Ag/AC nanocomposites, it is presumed that negatively charged bacteria (*E. coli* and total coliform), when come in contact with PPY-Ag nanocomposite/AC fibers, become dead via charge neutralization mechanism. Antistatic behavior of conducting polymer also supports the antimicrobial performance of composite coated membrane. However, most likely due to a synergistic effect of the cationic PPY ring and Ag nanoparticles, 100% reduction in bacteria from wastewater was finally obtained. Moreover, the bactericidal performance of PPY-Ag nanocomposites increased as the amount of loaded silver increased. In addition, large surface area of the nanomaterials may also contribute to the excellent bactericidal activity of PPY-Ag nanocomposites.

9.6 Conclusion

This chapter describes the development of PPY/Ag-NPs/AC nanocomposite–based membrane for the removal of *E. coli* and total coliform from wastewater. Encapsulation of Ag-NPs into the PPY matrix has been confirmed by FTIR, XRD, and morphological characterization. Nanocomposites showed spherical morphology with diameter sizes ranged between 20 nm and 50 nm. Importantly, the synthesized PPY-Ag nanocomposites had an excellent biocidal potential against *E. coli* and total coliform bacteria. The most important requirement of water filtration membranes is total reduction of bacteria from wastewater. PPY-Ag nanocomposite based membrane found to be efficient for this requirement. The output count of *E. coli* and total coliform in the effluent was below the detection limit (<10 CFU/100 ml).

Efficiency of nanocomposite-coated membrane was evaluated by flowing about 30 L wastewater through the membrane and results indicated that stronger antibacterial activity of PPY-Ag-impregnated AC may be a result of electrostatic interactions. Bacterial strains such as *E. coli* and total coliform with an extracellular capsule carry less negative charge and are less prone to adsorption to the positively charged surface of conducting polymer–based nanocomposites. Overall, the findings suggest the use of PPY-Ag-NPs/AC membrane as an effective and sustainable point-of-use water treatment technology. The use of such a conducting polymer nanocomposites impregnated with AC-based membrane may support the production of water free from coliform bacteria (i.e., *E. coli* and total coliform) and other contaminants. The synthesized PPY-Ag nanocomposite–based fiber may also find applications in other fields such as antistatic materials, bioadhesives, the coating of biomedical materials, clinical wound dressings, and as a water filter for wastewater treatment.

References

1. Krasner S. W., Weinberg H. S., Richardson S., Pastor S. J., Chin R., Sclimenti M. J., Onstad G. D., Thruston A. D. 2006. *J. Environ. Sci. Technol.*, 40, 7175–7185.
2. Srivastava A., Srivastava O. N., Talpatra S., Vijtai R., Ajayan P. M. 2004. *Nature Mat.*, 3, 610–614.
3. Kang S., Pinault M., Pfefferle L. D., Elemelech M. 2007. *Langmuir*, 23, 8670–8673.
4. Qi L., Xu Z., Jiang X., Hu C., Zou X. 2004. *Carbohydr. Res.*, 339, 2693–2700.
5. Cho M., Chung H., Choi W., Yoon J. 2005. *Appl. Environ. Microbiol.*, 71, 270–275.
6. Kim J. Y., Park C., Yoon J. 2008. *Environ. Eng. Res.*, 13, 36–140.
7. Lyon D. Y., Adams L. K., Falkner J. C., Alvarez P. J. J. 2006. *Environ. Sci. Technol.*, 40, 4360–4366.
8. Jones N., Ray B., Ranjit K. T., Manna A. C. 2008. *FEMS Microbio. Lett.*, 279, 71–76.
9. Zvekic D., Srdic V. V., Karaman M. A., Matavulj M. N. 2011. *Process. Appl. Cera.*, 5, 41–45.

10. Singh G., Joyce E. M., Beddow J., Mason T. J. 2012. *J. Microbiol. Biotechnol. Food Sci.*, 2, 106–120.
11. Morones J. R., Elechiguerra J. L., Camacho A., Holt K., Kouri J. B., Ramirez J. T., Yacaman M. J. 2005. *Nanotechnology*, 16, 2346–2353.
12. Sheikh N., Akhavan A., Kassaee M. Z. 2009. *Physica E*, 42, 132–135.
13. Dror-Ehre A., Mamane H., Belenkova T., Markovich G., Adin A. 2009. *J. Colloid. Interf. Sci.*, 339, 521–526.
14. Li Q., Mahendra S., Lyon D. Y., Brunet L., Liga M. V., Li D., Alvarez P. J. J. 2008. *Water Res.*, 42, 4591–4602.
15. Le Pape H., Solano-Serena F., Contini P., Devillers C., Maftah A., Leprata P. 2002. *Carbon*, 40, 2947–2954.
16. Tamai H., Katsu N., Ono K., Yasuda H. 2001. *Carbon*, 39, 1963–1969.
17. Oya A., Wakahara T., Yoshida S. 1993. *Carbon*, 31, 1243–1247.
18. Wang Y. L., Wan Y. Z., Dong X. H., Cheng G. X., Tao H. M., Wen T. Y. 1998. *Carbon*, 36, 1567–1571.
19. Yamamoto O., Sawai J., Sasamoto T. 2002. *Mat. Trans.*, 43, 1069–1073.
20. Zodrow K., Bronet L., Mahendra S., Li D., Zhang A., Li Q., Alvarez P. J. J. 2008. *Water Res.*, 43, 715–723.
21. Shanmugam S., Viswanathan B., Varadarajan T. K. 2006. *Mat. Chem. Phys.*, 95, 51–55.
22. Botes M., Cloete T. E. 2010. *Crit. Rev. Microbiol.*, 36, 68–81.
23. Cho K. H., Park J. E., Osaka T., Park S. G. 2005. *Electrochimica Acta*, 51, 956–960.
24. Sondi I., Sondi B. S. 2004. *J. Colloid. Interf. Sci.*, 275, 177–182.
25. Shiraishi Y., Toshima N. 2000. *Colloid. Surf. A: Physiochem. Eng. Asp.*, 169, 59–66.
26. Olivera M. M., Castro E. G., Canestraro C. D., Zanchet D., Ugarte D., Roman L. S., Zarbin A. J. G. 2006. *J. Phys. Chem. B*, 110, 17063–17069.
27. Schierholz J. M., Lucas L. J., Rump A., Pulverer G. 1998. *J. Hospital Infection*, 40, 257–262.
28. Betancourt-Galindo R., Cabrera Miranda C., Puente Urbina B. A., Castañeda-Facio A., Sanchez-Valdes S., Mata Padilla J., García Cerda L. A., Perera Y. A., Rodriguez-Fernandez O. S. 2012. *Nanotechnology*, 2012, doi:10.5402/2012/186851.
29. Malinauskas A. 2001. *Polymer*, 42, 2957–2972.
30. Seshadri D. T., Bhat N. V. 2005. *Ind. J. Fib. Tex. Res.*, 30, 204–206.
31. Seshadri D. T., Bhat N. V. 2005. *Sen'i Gakkaishi*, 61, 104–109.

32. Zhang S., Shao Y., Liu J., Aksay I. A., Lin Y. 2011. *ACS Appl. Mater. Interf.*, 3, 3633–3637.
33. Varesano A., Aluigi A., Florio L., Fabris R. 2009. *Syn. Met.*, 159, 1082–1089.
34. Wu C. S. 2011. *Polym. Int.*, 60, 807–815.
35. Wu C. S. 2012. *Exp. Polym. Lett.*, 6, 465–475.
36. Li C. Y., Wan Y. Z., Wang J., Wang Y. L., Jiang X. Q., Han L. M. 1998. *Carbon*, 36, 61–65.
37. Nwabanne J. T., Igbokwe P. K. 2012. *Int. J.of App.Sci. and Technol.*, 2 (5):106- 115.
38. Grabow W. O. K., du Preez M. 1979. *Appl. Environ. Microbiol.*, 38, 351–358.
39. Fifield C. W., Schaufus C. P. 1958. *J. Am. Water Works Assoc.*, 50, 193–196.
40. Eaton A. D., Clesceri L. S., Greenberg A. E. 1998. *Standard Methods for the Examination of Water and Wastewater*, 20th ed., American Public Health Association, Washington, D.C.
41. Varshney S., Singh K., Ohlan A., Jain V. K., Dutta V. P., Dhawan S. K. 2012. *J. Alloys Comp.*, 538, 107–114.
42. Silverstein R. M., Webster F. X. 2002. *Spectrometric Identification of Organic Compounds*, 6th edition, John Wiley and Sons, New Delhi.
43. Allena N. S., Murray K. S., Fleming R. J., Saunders B. R. 1997. *Synth. Met.*, 87, 237.
44. Pinter E., Patakfalvi R., Fulei T., Gingl Z., Dekany I., Visy C. 2005. *J. Phys. Chem. B*, 109, 17474–17478.
45. Chubb J. N. 2002. *J. Electrostatics*, 54, 233.
46. Bhandari H., Bansal V., Choudhary V., Dhawan S. K. 2009. *Polym. Inter.*, 58, 489.
47. Bhandari H. 2011. *Synthesis, Characterization and Evaluation of Conducting Copolymers for Corrosion Inhibition and Antistatic Applications*, PhD Thesis, IIT (2006PTZ8011).
48. Khan M. A. M., Kumar S., Ahamed M., Alrokayan S. A., Al-Salhi M. S. 2011. *Nano. Res. Lett.*, 6, 434.
49. Sharma V. K., Yngard R. A., Lin Y. 2009. *Adv. Colloid. Interf. Sci.*, 145, 83–96.
50. Shrivastava S., Bera T., Roy A., Singh G., Ramachandrarao P., Dash D. 2007. *Nanotechnology*, 18, 1–9.
51. Pal S., Tak Y. K., Song J. M. 2007. *App. Environ. Microbiol.*, 73, 1712–1720.

Chapter 10

Titanium Dioxide–Based Materials for Photocatalytic Conversion of Water Pollutants

Sónia A. C. Carabineiro,[a] Adrián M. T. Silva,[a] Cláudia G. Silva,[a] Ricardo A. Segundo,[a] Goran Dražić,[b,c] José L. Figueiredo,[a] and Joaquim L. Faria[a]

[a]*Laboratory of Catalysis and Materials, Associate Laboratory LSRE/LCM, Faculdade de Engenharia, Universidade do Porto, Rua Dr. Roberto Frias, 4200-465 Porto, Portugal*
[b]*Jozef Stefan Institute, Department of Nanostructured Materials, Jamova 39, SI-1000 Ljubljana, Slovenia*
[c]*Laboratory for Materials Chemistry, National Institute of Chemistry, Ljubljana, Slovenia*
scarabin@fe.up.pt

Heterogeneous photocatalysis has proven its value towards meeting the challenges posed by cost-effective treatments of wastewater. In spite of the enormous number of pollutants that have been the subject of photocatalytic treatment studies, knowledge on the photoinitiated advanced oxidation processes for the conversion of hazardous emerging chemical substrates introduced to the water ecosystem remains incomplete. Many questions concerning the complex and interrelated chemical and photochemical mechanisms of conversion still need to be answered, limiting the effectiveness

Nanocomposites in Wastewater Treatment
Edited by Ajay Kumar Mishra
Copyright © 2015 Pan Stanford Publishing Pte. Ltd.
ISBN 978-981-4463-54-6 (Hardcover), 978-981-4463-55-3 (eBook)
www.panstanford.com

of the process. A nanotechnological approach can be used to harness the photocatalytic capabilities of titanium dioxide (TiO_2), the most used catalyst in this type of applications, generating nanoarchitectures with extremely high surface area and remarkable redox properties, much more efficient than the ordinary grades of the oxide. In this chapter, we will describe some of our recent results on the degradation of critical and emerging pollutants present in wastewater, such as diphenhydramine and phenolic compounds, by using TiO_2 alone or loaded with gold nanoparticles.

10.1 Introduction

Water is essential for life. It is present in several human activities and is needed for drinking (a large percentage of the human body is water), food preparation, cleaning, agriculture, industry, and energy production. The demand for fresh water has increased with population growth and with intensification of agricultural and industrial activities. Therefore, it is crucial to protect the natural sources of water from pollutants and to develop new technologies for wastewater treatment. Biological methods are simple, but they produce sludge and occupy large areas. Chemical treatment can efficiently convert or mineralize the pollutants. However, in the presence of highly stable compounds, it may be needed to use more reactive systems.

Heterogeneous photocatalysis is one of the most promising technologies for water treatment [1–6]. It uses oxygen as oxidizing agent, requires mild operation conditions of temperature and pressure, can be done simply by using sunlight, making it a cheap process. It also requires a semiconductor photocatalyst. Titanium dioxide (TiO_2) is the material most frequently used for this purpose, due to its high photocatalytic activity, chemical stability, and low cost [3, 6–11]. It exists mainly in three polymorphic forms: anatase, rutile, and brookite, with crystal structures displayed in Fig. 10.1. Rutile (Fig. 10.1b) is the most stable form of TiO_2. Anatase (Fig. 10.1a) and brookite (Fig. 10.1c) are metastable and can be readily transformed into rutile upon heating. The crystalline structures consist of TiO_2^{6-} octahedra arranged in a tetragonal unit cell geometry [9, 12]. The crystal structures differ by the distortion of each octahedron and the assembly patterns of the octahedral chains. The octahedrons are coupled by the vertices in anatase, by the edges in rutile, and by both vertices or edges in brookite (Fig. 10.1).

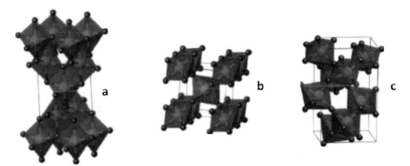

Figure 10.1 Crystal structures of anatase (a), rutile (b), and brookite (c) (adapted from [13]).

The synthesis of TiO_2 is commonly achieved by the sol-gel method, involving the hydrolysis of titanium chlorides or alkoxides [6, 14–20]. A liquid suspension of solid particles smaller than 1 µm (sol) is obtained by hydrolysis and partial condensation of the alkoxide. A gel is formed by condensation of the sol particles into a three-dimensional network. The phases normally obtained by this sol-gel synthesis are anatase and rutile (along with some amorphous titanate), depending on the calcination temperature [6, 18, 19].

Already in 1972, Fujishima and Honda showed the possibility of water splitting by a photoelectrochemical cell with a rutile TiO_2 photoanode and a Pt counter electrode [21]. Soon after, Carey et al. first reported the photocatalytic degradation of biphenyl and chlorobiphenyl molecular derivatives using TiO_2 [22]. Afterwards, several additional reports were published, showing the versatility of TiO_2 in photocatalysis for water purification [3, 6, 7, 9, 15, 17, 18]. However, TiO_2 has limited activity under visible light irradiation, which motivated the quest for modified TiO_2 materials absorbing visible light. Doping with metals, such as Au, has proven to be successful in enhancing photocatalytic activity of TiO_2 materials [23–35]. Gold is a noble metal that does not undergo photocorrosion and can be strongly anchored on TiO_2 surface, exhibiting a characteristic surface plasmon band in the visible region, due to the collective excitation of electrons in the gold nanoparticles [13, 36].

This chapter describes some recent results on the degradation of critical and emerging pollutants present in wastewater—such as diphenhydramine, pharmaceutical and phenolic compounds such as mono- and di-nitrophenols, and *para*-substituted phenols such as 4-chlorophenol, 4-cyanophenol, and 4-methoxyphenol—using TiO_2 alone or loaded with gold nanoparticles as photocatalysts.

Phenolic compounds are poisonous benzene derivatives that contaminate industrial wastewaters [6, 18, 37–43], being refractory and recalcitrant species in conventional biological treatment processes. These pollutants are highly stable, toxic, and carcinogenic, capable of causing considerable damage and threat to the ecosystem and human health [6].

Diphenhydramine (DP) hydrochloride, that is, 2-(diphenylmethoxy)-N,N-dimethylethylamine hydrochloride, is the active ingredient of Benadryl®, an antihistaminic drug that combines sedative, antiemetic, antitussive, and hypnotic properties, used in the treatment of allergies, allergic rhinitis, hives, itching, and insect bites and stings [44]. The particular persistence of DP in natural waters is mainly related to its low biodegradability, and it has also shown high toxicity with mutagenic and carcinogenic effects [45].

10.2 Experiments

10.2.1 Preparation of Titanium Dioxide Supports

Two types of TiO_2 were used in this study: commercial TiO_2 from Degussa (Evonik), also known as P25, and TiO_2 synthesized [6, 18–20] at room temperature, through the addition of 0.1 mol (19.2 mL) of titanium isopropoxide [$Ti(OC_3H_7)_4$, Aldrich 97%] to 125 mL of ethanol (C_2H_5OH, Panreac 99%). This solution was stirred for 30 min, and then 1 mL of nitric acid (HNO_3, Panreac 65%) was added. The mixture was kept under constant stirring until a homogeneous gel was formed. The latter was allowed to age in air for 1 week before being ground into a fine powder ($d < 0.1$ mm). This powder was calcined at different temperatures, between 400 °C and 700 °C, in a vertical tubular furnace, under a constant nitrogen flow (2 h, 75 cm^3 min^{-1}), in order to obtain different compositions of anatase and rutile (results shown in Table 10.1), and it was found that calcination at 500 °C provided a sample with similar amounts of anatase and rutile close to that of the commercial P25 [6, 18–20]. Therefore, the sample calcined at 500 °C (TiO$_{2sg500}$) was also used for comparison purposes.

10.2.2 Gold Loading

Gold loading was carried out by double impregnation (DIM), liquid phase reductive deposition (LPRD), photodeposition (PD), and ultrasonication (US) methods. The DIM method is similar to traditional impregnation (the support is impregnated with a solution of HAuCl$_4$ using sonication), but using a second impregnation step with addition of an aqueous solution of Na$_2$CO$_3$ (1 M), under constant ultrasonic stirring [46–50]. LPRD consists of mixing a solution of the gold precursor with a solution of NaOH (1:4 by weight) with stirring at room temperature. The resulting solution is aged for 24 h in the dark at room temperature, to complete the hydroxylation of Au^{3+} ions. Then the appropriate amount of support is added to the solution and, after ultrasonic dispersion for 30 min, the suspension is aged in the oven at ~100 °C overnight [48–51]. For PD, the Au precursor is dissolved in water and methanol (300 mL, 15:1 ratio), mixed with the support, sonicated for 30 min to improve dispersion, and photodeposited using a UV lamp (Heraeus TNN 15/32) with an emission line at 253.7 nm (ca. 3 W of radiant flux) [46, 52]. A series of tests were carried out at different pH values (from 1 to 12, not buffered) before photodeposition. Also different photodeposition times (1–4 h) were used. US consists in dissolving the gold precursor in water and methanol (15:1 by weight), mixing with the support, and sonicating for 8 h [46]. In all cases, the resulting solid is washed repeatedly with distilled water for chloride removal (which is well known to cause sintering of Au nanoparticles, thus turning them inactive [51, 53–55]), and dried in the oven at ~110 °C overnight.

10.2.3 Characterization Techniques

The materials were characterized by the adsorption of N$_2$ at −196 °C, in a Quantachrome NOVA 4200e apparatus. X-ray diffraction (XRD) analysis was carried out in a PANalytical X'Pert MPD equipped with X'Celerator detector and secondary monochromator

(Cu Kα = 0.154 nm, 50 kV, 40 mA). The collected spectra were analyzed by Rietveld refinement using Powder Cell software, allowing the determination of the grain size.

Transmission electron microscopy (TEM) experiments for the TiO_2 samples were carried out in a JEOL 1011 transmission electron microscope operating at 100 kV. The samples were dispersed in ethanol using an ultrasonic bath for 1 min. A drop of the dispersion was placed on a holey carbon film deposited on a copper grid and left to dry at ambient temperature. A JEOL 2010F analytical electron microscope, equipped with a field-emission gun, was used for high-resolution transmission electron microscopy (HRTEM) investigations for Au/TiO_2 materials. The microscope was operated at 200 keV, and an energy dispersive X-ray spectrometer (EDXS) LINK ISIS-300 from Oxford Instruments with a UTW Si–Li detector employed for the chemical analysis. The samples for TEM were prepared from a diluted suspension of nanoparticles in ethanol. A drop of suspension was placed on lacey carbon-coated Ni grid and allowed to dry in air. Z-contrast images were collected using a high-angle annular dark-field detector (HAADF) in scanning transmission mode (STEM). Particle sizes were measured from TEM images, using Image J program. Histograms were drawn from measurements made on 100–500 particles, depending on the sample. Average particle sizes were calculated for all samples.

The UV–Vis spectra of the TiO_2 materials were measured on a JASCO V-560 UV–Vis spectrophotometer, equipped with an integrating sphere attachment (JASCO ISV-469). The spectra were recorded in diffuse reflectance mode and transformed by the instrument software (JASCO) to equivalent absorption Kubelka–Munk units.

10.2.4 Catalytic Tests

The photocatalytic degradation studies of DP were carried out at room temperature, in a glass cylindrical microreactor filled with 7.5 mL of a 10 or 100 mg L^{-1} DP aqueous solution. The catalyst load was kept at 1 g L^{-1}. The solution was magnetically stirred and continuously purged with an oxygen flow. The irradiation source consisted in a Heraeus TQ 150 medium-pressure mercury vapor lamp (λ_{exc} = 254, 313, 365, 436, and 546 nm). A DURAN® glass-cooling jacket was used for irradiation in the near-UV to visible light range (λ_{exc} = 365, 436, and 546 nm). The lamp was located 7.5 cm from the reactor. To test the materials only with visible light, a long-pass filter with a cut edge at 450 nm was used (all electromagnetic radiation with λ < 430 nm was blocked).

In the case of the photocatalytic reactions with phenolic compounds, a glass-immersion photochemical reactor charged with 250 mL of a 100 mg L^{-1} solution of each compound was used, instead of a microreactor. Before turning illumination on, the suspension was magnetically stirred in the dark for 30 min to establish an adsorption–desorption equilibrium. The first sample was taken out at the end of the dark adsorption period, just before the light was turned on, to determine the concentration in solution, which was hereafter considered the initial concentration (C_0) after dark adsorption. The DP concentration was followed by high performance liquid chromatography (HPLC) using a Hitachi Elite LaChrom apparatus equipped with an L-2450 diode array detector and a solvent delivery pump L-2130. An isocratic method (flow rate of 1 mL min^{-1}) was used with the eluent consisting of an A:B (70:30) mixture of 20 mM NaH$_2$PO$_4$ acidified with H$_3$PO$_4$ at pH = 2.80 (A) and acetonitrile (B). The absorbance was found to be linear over the whole range considered. In some samples, the total amount of organic carbon (TOC) was measured at 60 min. To test the reaction mechanisms, some scavengers of HO• radicals and holes were used, namely, *tert*-butanol (t-but) and ethylenediaminetetraacetic acid (EDTA).

10.3 Results and Discussion

10.3.1 Characterization of TiO$_2$ Materials

The XRD results showed that the crystal phase of P25 is a mixture of 81% anatase and 19% rutile (Fig. 10.2c and Table 10.1). Concerning the sol-gel samples, at temperatures below 450 °C, the only crystal phase detected was anatase (Fig. 10.2a and Table 10.1). Transformation of anatase to rutile occurs between 450 °C and 500 °C. Above 500 °C, there is a mixture of anatase and rutile phases, with different proportions, according to the calcination temperature. The higher the temperature, the larger the amount of rutile; this being the only phase at 700 °C because anatase is totally converted.

Table 10.1 Surface area (S), crystallite dimensions of anatase (d_A) and rutile (d_R), content of anatase (X_A) and rutile (X_R) for the commercial P25 sample and for TiO$_2$ samples obtained by the sol-gel method at different calcination temperatures

Sample	S (m²/g)	d_A (nm)	d_R (nm)	X_A (%)	X_R (%)
TiO$_2$ P25	51	21	28	81	19
TiO$_{2sg400}$	107	8.5	—	100	—
TiO$_{2sg450}$	84	10	—	100	—
TiO$_{2sg500}$	26	14	20	69	31
TiO$_{2sg550}$	13	23	24	38	62
TiO$_{2sg600}$	3	37	39	5	95
TiO$_{2sg700}$	3	—	39	—	100

Source: Adapted from [18].

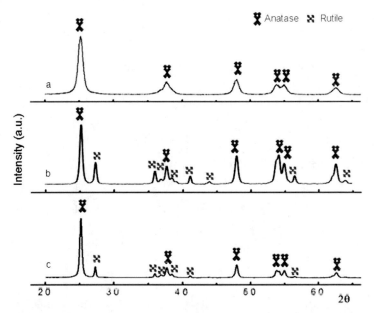

Figure 10.2 XRD patterns of TiO$_2$ sol-gel calcined at 400 °C (a) and at 500 °C (b) and that of TiO$_2$ P25 (c).

The crystallite sizes (*d*) and the corresponding phase percentages (*X*) are also listed in Table 10.1. The crystallite dimensions were calculated from the broadening of the (101) reflection for anatase and (110) reflection for rutile [6], using Scherrer's equation

$$d = \frac{k\lambda}{\beta \cos\theta},$$

where *k* is a constant that depends on the crystallite shape (0.9, with the assumption of spherical particles), λ is the X-ray wavelength, β is the full width at half maximum of the selected peak, and θ is the Bragg diffraction angle of the peak. The rutile-to-anatase ratio was determined from the relation [56]

$$X_A = \left(1 + 1.26 \frac{I_R}{I_A}\right)^{-1}$$

where X_A is the anatase fraction in the catalyst, and I_R and I_A are the integral intensities of the (110) and (101) peaks of rutile and anatase, respectively.

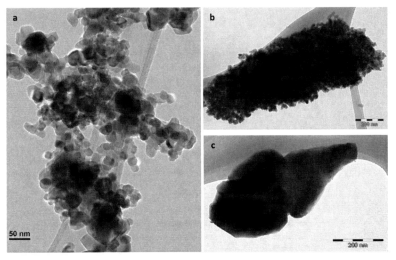

Figure 10.3 TEM micrographs of TiO$_2$ P25 —anatase and rutile (a)— and samples obtained by sol-gel calcined at 400 °C —anatase only (b)— and 700 °C —rutile only (c).

The XRD diffraction peaks were relatively broad for samples calcined at low temperatures, especially at 400 °C (Fig. 10.2a). This may be due to the small crystallite size and the presence of amorphous (noncrystallized) TiO$_2$. Increasing the calcination temperature increases the crystalline size of both anatase and rutile (Table 10.1).

The characterization results obtained by N$_2$ adsorption at −196 °C are also displayed in Table 10.1. It can be seen that the surface area decreases with increasing calcination temperature for the sol-gel samples. This is related to the increase in the crystallite dimensions, as confirmed by XRD.

The TEM analysis of TiO$_{2sg400}$ and TiO$_{2sg700}$ samples (Fig. 10.3) showed crystallites with sizes similar to those determined by XRD, as expected. P25 showed a mixture of particles with different sizes (Fig. 10.3a), while smaller anatase particles were found for TiO$_{2sg400}$ (Fig. 10.3b) and larger rutile particles for TiO$_{2sg700}$ (Fig. 10.3c). Higher temperature enhances the phase transformation from thermodynamically metastable anatase to the more stable and condensed rutile phase.

10.3.2 Characterization of Au/TiO$_2$ Materials

Selected Au/TiO$_2$ samples were analyzed by HRTEM and HAADF (Fig. 10.4). The presence of gold was confirmed by EDXS (Fig. 10.4d) and shown to be ~0.9%, therefore close to the intended loading (1% wt.). Figure 10.4a,b shows HAADF and HRTEM images of TiO$_2$ P25 with gold loaded by LPRD and PD, respectively. In the HAADF images, the gold nanoparticles are seen as bright spots (Fig. 10.4a), but as darker spots in HRTEM images (Fig. 10.4b,c), as confirmed by EDXS (Fig. 10.4d).

Table 10.2 contains a summary of the gold nanoparticle sizes and ranges obtained. The Au nanoparticle size distributions for P25 with Au loaded by different methods are shown in Fig. 10.5. LPRD has lower average size (Table 10.2 and Fig. 10.5a), these being the smallest gold nanoparticles found in this study. PD performed at pH 10, DIM, and US (Fig. 10.5 and Table 10.2) produced larger sizes.

Table 10.2 Au nanoparticle average size and range for the Au/TiO$_2$ samples prepared by different methods

Sample	Au average size (nm)	Au size range (nm)
TiO$_{2sg500}$ Au LPRD	7.9	2–15
TiO$_2$ P25 Au LPRD	3.9	1–11
TiO$_2$ P25 Au PD pH10	6.5	3–14
TiO$_2$ P25 Au DIM	9.3	4–16
TiO$_2$ P25 Au US	15.1	5–25

Diffuse reflectance UV-Vis spectra of Au-loaded P25 are shown in Fig. 10.6. As expected, due to the presence of gold nanoparticles, they exhibit two main bands: one in the visible region between 500 nm and 650 nm typical of the gold surface plasmon band, and a stronger band in the UV region, with an edge rising at 400 nm, which is due to the band gap transition of TiO$_2$ semiconductor.

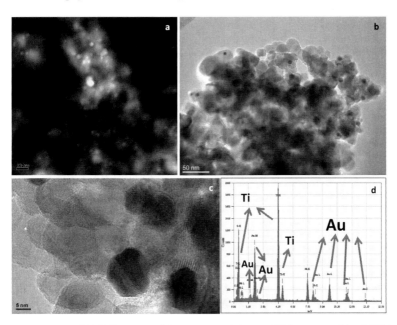

Figure 10.4 HAADF image of Au nanoparticles on P25 prepared by LPRD (a), seen as bright spots, and TEM images of Au nanoparticles on P25 prepared by PD (b) and on TiO$_{2sg500}$ prepared by LPRD (c) with EDXS spectrum of a gold nanoparticle (d).

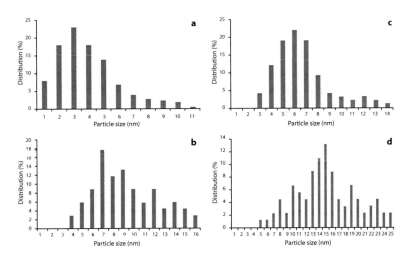

Figure 10.5 Size distribution histograms of Au nanoparticles on P25 prepared by LPRD (a), PD at pH 10 (b), DIM (c), and US (d).

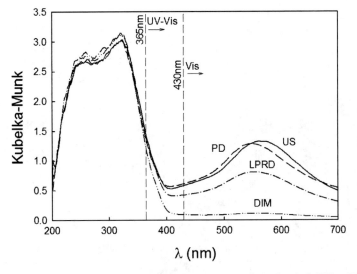

Figure 10.6 Diffuse reflectance UV-Vis spectra of Au-loaded TiO$_2$ P25 prepared by different techniques.

10.3.3 Catalytic Results for DP Photodegradation

Au/TiO$_2$ P25 materials were tested for DP photodegradation. Regardless of the wavelength range of irradiation (UV-Vis or

Vis), addition of Au by US did not produce a positive effect in the photoactivity of TiO$_2$, which may be related to the larger dimensions of the Au nanoparticles (Fig. 10.7). In UV-Vis conditions ($\lambda \geq 365$ nm), addition of Au by PD has a moderate improvement on the activity of P25 catalyst (Fig. 10.7a). However, when using visible light radiation ($\lambda > 430$ nm), addition of Au by LPRD produced a large improvement on the activity of P25 (Fig. 10.7b).

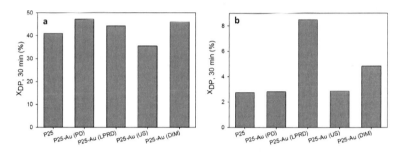

Figure 10.7 DP (100 mg L^{-1}) conversion at the end of 30 min of (a) UV-Vis irradiation ($\lambda \geq 365$ nm); (b) Vis irradiation ($\lambda > 430$ nm).

A different mechanism must be considered depending on the wavelength of radiation used in the photocatalytic experiments. For UV radiation, the most reasonable rationalization of the photocatalytic mechanism assumes direct photoexcitation of TiO$_2$, leading to the generation of electrons in the semiconductor conduction band and electron holes in the valence band. The electrons in the conduction band will move to the gold nanoparticles acting as electron buffer (Fig. 10.8a)[57]. This is expected to be the main mechanism when using near UV-Vis ($\lambda \geq 365$ nm) radiation, with the more energetic part of the radiation used (between 365 nm and 400 nm) being absorbed by TiO$_2$. On the other hand, when irradiating at $\lambda > 430$ nm (Vis), no absorption by TiO$_2$ is expected and a distinct mechanism is proposed in which, upon photoexcitation of Au nanoparticles, electrons from Au are injected into the TiO$_2$ conduction band leading to the generation of holes in the Au nanoparticles and electrons in the TiO$_2$ conduction band (Fig. 10.8b) [57]. As can be observed in Fig. 10.6, gold nanoparticles supported in TiO$_2$ exhibit a surface plasmon band with λ_{max} at about 550 nm, very close to one of the excitation lines used in these experiments ($\lambda_{exc} = 546$ nm). This special feature is related to the very small dimensions of gold nanoparticles. As in other heterogeneous catalytic systems, it is well known that the nanoparticle size of gold is related with its catalytic activity [52, 54, 55, 58]. In fact, the catalyst prepared by LPRD presents the smallest Au nanoparticles (Table 10.2 and Fig. 10.5), which can explain the highest photocatalytic activity under visible light irradiation.

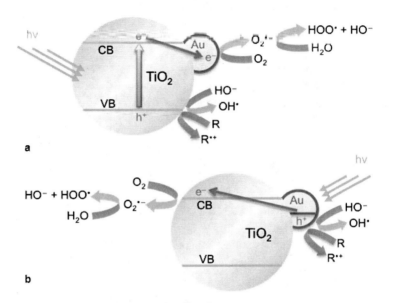

Figure 10.8 Proposed rationalization of the photocatalytic activity of Au/TiO$_2$ under UV (a) or Vis (b) light excitation.

To better understand the oxidation/reduction pathways occurring at the surface of the catalyst during the photocatalytic degradation of DP, EDTA and *t*-BuOH were added to the system as hole and radical scavengers, respectively (Fig. 10.9). Results show that the presence of *t*-BuOH as radical scavenger slows down the photocatalytic degradation of DP, but the addition of EDTA reduced the DP photodegradation rate to a much higher extent. It is, therefore, clear that reactive species such as hydroxyl (HO•), superoxide (O$_2$•−), and hydroperoxyl (HOO•) radicals (or even singlet oxygen, ^{1}O$_2$) participate in the photocatalytic mechanism under near UV-Vis or only Vis irradiation, but the photogenerated holes play a major role in the degradation mechanism.

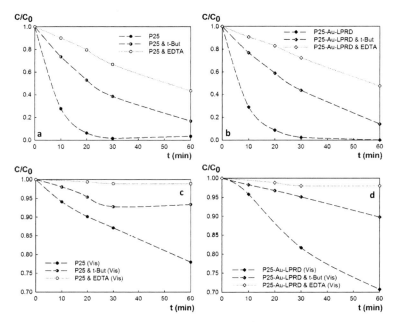

Figure 10.9 Photodegradation of DP (10 mg L^{-1} in the presence of t-but or EDTA) using P25 with and without Au—UV-Vis results (a) and (b); Vis results (c) and (d).

It was observed that addition of Au to the TiO$_{2sg500}$ sample did not improve the catalyst activity (results not shown), except when Au was loaded by LPRD (in this case, only a slight improvement was observed in relation to the support). The reason why supports with similar contents of anatase and rutile show different catalytic behaviors is likely due to the unique features of the P25 material, namely, the much lower particle size (of the order of several nanometers compared to around 100 μm in the case of the sol-gel TiO$_2$), leading to a better dispersion in aqueous media, as well as a special distribution of both anatase and rutile phases, which increases the efficiency of the electron–hole separation process [9, 59].

10.3.4 Photocatalytic Degradation of Phenolic Compounds using P25 Catalyst

It is known that the photocatalytic reactivity of aromatic compounds can be affected by the number of substituents, their electronic nature, and their position on the aromatic ring. Indeed, the photocatalytic degradability of phenolic compounds has been correlated with the nature of the different substituents present in the aromatic system [38, 60].

Photocatalytic degradation of mono-substituted (2-, 3-, and 4-nitrophenol, hereafter referred as 2-NP, 3-NP, and 4-NP, respectively) and di-substituted (2,4- and 2,6-dinitrophenol, hereafter referred as 2,4-DNP and 2,6-DNP, respectively) nitrophenols was performed using P25 as catalyst under near UV-Vis light irradiation. All the reactions followed a pseudo-first-order kinetic model. Rate constants obtained for the reactions with each molecule are shown in Fig. 10.10.

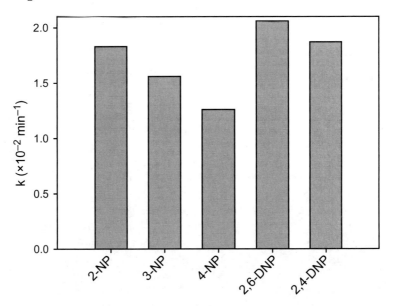

Figure 10.10 First-order kinetic rate constants obtained for the photocatalytic degradation reactions of 2-NP, 3-NP, 4-NP, 2,4-DNP, and 2,6-DNP.

Photocatalytic reactions are recognized as electrophilic processes, being accelerated by the presence of electron-donating (activating) groups in aromatic molecules and retarded when electron-withdrawing (deactivating) substituents are present. Additionally, when an aromatic ring has more than one functional group, the effects of the substituents are combined and their total effect is generally the result of the sum of these different contributions. The hydroxyl group (–OH) is an electron-donating substituent, thus activating the ring for an electrophilic attack. Since orientation in aromatic substitution is normally a consequence of kinetic control, the hydroxyl is said to be an *ortho–para* director. On the other hand, nitro (–NO$_2$) is an electron-withdrawing group, with a powerful deactivating action, thus a *meta*-director. Therefore, one can conclude that the aromatic ring will suffer the effect of both groups: increase in electron density by the hydroxyl groups and simultaneously removal of electron density by the nitro groups. The later effect should be more intense, taking into consideration the influence of the substituents as measured by their dipole moment (1.45 D for –OH as activating and 3.97 D for –NO$_2$ as deactivating substituent of electrophilic attack) [61]. In this context, the carbons connected to the nitro substituents will be electron rich, therefore more prone to attack by the hydroxyl radical, resulting in HO•-induced denitration, which is in accordance with the obtained results, i.e., the reactivity of the phenolic compounds increases with the number of nitro groups attached to the phenolic ring (higher k values for di-substituted nitrophenols).

Comparing the reactivity found for mono-nitrophenols, an *ortho* nitro substituent (2-NP) makes the molecule more reactive than a *meta* (3-NP) or *para* nitro group (4-NP). This result indicates that the free hydroxyl radical should be preferentially directed to the attack when both hydroxyl and nitro substituents are as near as possible, resulting in a higher electron density at the *ortho*-position with respect to –OH.

To study the effect of the presence of different substituents on their photocatalytic reactivity, four *para*-substituted phenols were used, namely, 4-chlorophenol, 4-cyanophenol, 4-methoxyphenol, and 4-nitrophenol. The influence that a substituent has on the electronic character of a particular aromatic system is represented by the Hammett constant (σ). By definition, hydrogen *para*-substituted phenol is taken as reference, with $\sigma_p = 0$. Generally, for reactions involving phenols, a modified Hammett constant (σ_p^-) is used since the substituent may enter into some resonance with the reaction site in an electron-rich transition state. A positive value of σ_p^- indicates the presence of an electron-withdrawing (deactivating) group, and a negative value an electron-donating (activating) group. The Hammett constants for methoxy (–OCH$_3$), chloro (–Cl), cyano (–CN), and nitro (–NO$_2$) groups are –0.27, 0.23, 0.88, and 1.25, respectively [62].

The influence of the substituent nature on the photodegradation of the four *para*-substituted phenols was investigated by comparing their initial rates of degradation (r_0) and correlating them with the corresponding σ_p^- values (Fig. 10.11).

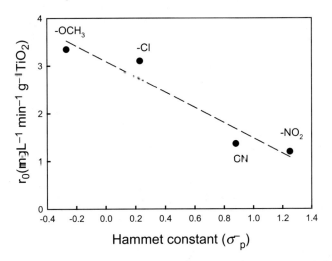

Figure 10.11 Relationship between initial degradation rates (r_0) and Hammett constants (σ_p^-) for the different *para*-substituted phenols.

Results show that the photocatalytic degradation rates depend to a good degree on the nature of the substituent, reactions being accelerated in the presence of electron-donating groups, and retarded by electron-withdrawing groups. The relationship found between the Hammett constants of the four *para*-substituted phenols and the respective initial reaction rates using both catalysts constitutes a useful tool to predict the behavior of these materials when used for the photocatalytic degradation of other phenolic molecules.

10.4 Conclusion

The most active samples were P25 with gold loaded by LPRD, followed by P25 with gold loaded by PD at pH 10. A basic value of pH was found to contribute to more active materials, possibly due to an increase in the hydroxylation of the Au precursor and elimination of chlorine, which is well known to contribute to the sintering and consequent inactivation of gold nanoparticles. P25 with Au loaded by LPRD was shown to be the most active catalyst for DP degradation, when visible light was used. The main mechanism of degradation for both catalysts and for both types of irradiation seems to be through the holes A TiO$_2$ prepared by the sol-gel method produced much worse results when compared with P25, when UV-Vis radiation was used. A slight improvement when gold was loaded was only seen for the LPRD sample, most likely as this method was the most adequate for gold deposition.

As for the phenolic compounds, the photocatalytic degradation rates depend on the nature of the substituent, reactions being accelerated in the presence of electron-donating groups, and retarded by electron-withdrawing groups. A relationship was found between the Hammett constants of the four *para*-substituted phenols and the respective initial reaction rates.

Acknowledgments

The authors acknowledge funding from Fundação para a Ciência e a Tecnologia (FCT) and FEDER in the framework of program COMPETE for PEst-C/EQB/LA0020/2013 and Project PTDC/EQU-ERQ/123045/2010, Program CIENCIA 2007 and Investigador FCT IF/01381/2013 grants (for SACC), Investigador FCT IF/01501/2013 grant (for AMTS) and Ph.D. grant SFRH/BD/74316/2010 (for RAS). This work was co-financed by QREN, ON3 and FEDER (Project NORTE-07-0124-FEDER-0000015). FCT and the Ministry of Higher Education, Science and Technology, Slovenia, are also acknowledged for financial support through the Portugal-Slovenia Cooperation in Science and Technology (2010/2011).

References

1. Fox M.A. and Dulay M.T. (1993). *Chemical Reviews*, 93, 341.
2. Hoffmann M.R., Martin S.T., Choi W., and Bahnemann D.W. (1995). *Chemical Reviews*, 95, 69.
3. Mills A. and Le Hunte S. (1997). *Journal of Photochemistry and Photobiology A: Chemistry*, 108, 1.
4. Herrmann J.M. (2005). *Topics in Catalysis*, 34, 49.
5. Fujishima A., Zhang X., and Tryk D.A. (2007). *International Journal of Hydrogen Energy*, 32, 2664.

6. Silva C.G. (2008). *Synthesis, Spectroscopy and Characterization of Titanium Dioxide Based Photocatalysts for the Degradative Oxidation of Organic Pollutants*, PhD Thesis, University of Porto.
7. Linsebigler A.L., Lu G., and Yates J.T. (1995). *Chemical Reviews*, 95, 735.
8. Diebold U. (2003). *Surface Science Reports*, 48, 53.
9. Carp O., Huisman C.L., and Reller A. (2004). *Progress in Solid State Chemistry*, 32, 33.
10. Kitano M., Matsuoka M., Ueshima M., and Anpo M. (2007). *Applied Catalysis A: General*, 325, 1.
11. Hwang K.-J., Lee J.-W., Shim W.-G., Jang H.D., Lee S.-I., and Yoo S.-J. (2012). *Advanced Powder Technology*, 23, 414.
12. Comotti M., Weidenthaler C., Li W.C., and Schuth F. (2007). *Topics in Catalysis*, 44, 275.
13. Landmann M., Rauls E., and Schmidt W.G. (2012). *Journal of Physics: Condensed Matter*, 24, 195503.
14. Moran-Pineda M., Castillo S., and Gomez R. (2002). *Reaction Kinetics and Catalysis Letters*, 76, 375.
15. Guo B., Liu Z., Hong L., and Jiang H. (2005). *Surface and Coatings Technology*, 198, 24.
16. Latt K.K. and Kobayashi T. (2008). *Ultrasonics Sonochemistry*, 15, 484.
17. Di Paola A., Cufalo G., Addamo M., Bellardita M., Campostrini R., Ischia M., Ceccato R., and Palmisano L. (2008). *Colloids and Surfaces A: Physicochemical and Engineering Aspects*, 317, 366.
18. Silva C.G. and Faria J.L. (2009). *Photochemical and Photobiological Sciences*, 8, 705.
19. Machado B.F. (2009). *Novel Catalytic Systems for the Selective Hydrogenation of Alpha-Beta Unsaturated Aldehydes*, University of Porto.
20. Gomes H.T., Machado B.F., Silva A.M.T., Drazic G., and Faria J.L. (2011). *Materials Letters*, 65, 966.
21. Fujishima A. and Honda K. (1972). *Nature*, 238, 37.
22. Carey J.H., Lawrence J., and Tosine H.M. (1976). *Bulletin of Environmental Contamination and Toxicology*, 16, 697.
23. Tian B.Z., Zhang J.L., Tong T.Z., and Chen F. (2008). *Applied Catalysis B-Environmental*, 79, 394.
24. Centeno M.A., Hidalgo M.C., Dominguez M.I., Navio J.A., and Odriozola J.A. (2008). *Catalysis Letters*, 123, 198.

25. Lu H.F., Zhou Y., Xu B.Q., Chen Y.F., and Liu H.Z. (2008). *Acta Physico-Chimica Sinica*, 24, 459.
26. Suprabha T., Roy H.G., and Mathew S. (2010). *Science of Advanced Materials*, 2, 107.
27. Yang D. and Lee S.-W. (2010). *Surface Review and Letters*, 17, 21.
28. Wang X. and Caruso R.A. (2011). *Journal of Materials Chemistry*, 21, 20.
29. Primo A., Corma A., and Garcia H. (2011). *Physical Chemistry Chemical Physics*, 13, 886.
30. Carneiro J.T., Savenije T.J., Moulijn J.A., and Mul G. (2011). *Journal of Photochemistry and Photobiology A: Chemistry*, 217, 326.
31. Chusaksri S., Lomda J., Saleepochn T., and Sutthivaiyakit P. (2011). *Journal of Hazardous Materials*, 190, 930.
32. Fu P. and Zhang P. (2011). *Thin Solid Films*, 519, 3480.
33. Hidalgo M.C., Murcia J.J., Navio J.A., and Colon G. (2011). *Applied Catalysis A: General,* 397, 112.
34. Oros-Ruiz S., Pedraza-Avella J.A., Guzman C., Quintana M., Moctezuma E., del Angel G., Gomez R., and Perez E. (2011). *Topics in Catalysis*, 54, 519.
35. Wang N., Tachikawa T., and Majima T. (2011). *Chemical Science*, 2, 891.
36. Silva A.M.T., Zilhão N.R., Segundo R.A., Azenha M., Fidalgo F., Silva A.F., Faria J.L., and Teixeira J. (2012). *Chemical Engineering Journal*, 184, 213.
37. Silva C.G. and Faria J.L. (2010). *Applied Catalysis B: Environmental*, 101, 81.
38. Silva C.G. and Faria J.L. (2010). *Chemsuschem*, 3, 609.
39. Silva C.G., Wang W.D., and Faria J.L. (2008). In *Advanced Materials Forum IV* (Marques A.T., Silva A.F., Baptista A.P.M., Sa C., Alves F., Malheiros L.F., and Vieira M., eds), Vol. 587–588, Trans Tech Publications Ltd, Stafa-Zurich, p. 849.
40. Wang W.D., Serp P., Kalck P., and Faria J.L. (2005). *Journal of Molecular Catalysis A: Chemical*, 235, 194.
41. Wang W.D., Serp P., Kalck P., and Faria J.L. (2005). *Applied Catalysis B: Environmental*, 56, 305.
42. Wang W.D., Serp P., Kalck P., Silva C.G., and Faria J.L. (2008). *Materials Research Bulletin*, 43, 958.
43. Wang W.D., Silva C.G., and Faria J.L. (2007). *Applied Catalysis B: Environmental*, 70, 470.

44. Pastrana-Martínez L.M., Faria J.L., Doña-Rodríguez J.M., Fernández-Rodríguez C., and Silva A.M.T. (2012). *Applied Catalysis B: Environmental*, 113–114, 221.
45. Kinney C.A., Furlong E.T., Werner S.L., and Cahill J.D. (2006). *Environmental Toxicology and Chemistry Letters*, 25, 317.
46. Carabineiro S.A.C., Machado B.F., Bacsa R.R., Serp P., Drazic G., Faria J.L., and Figueiredo J.L. (2010). *Journal of Catalysis*, 273, 191.
47. Carabineiro S.A.C., Silva A.M.T., Drazic G., Tavares P.B., and Figueiredo J.L. (2010). *Catalysis Today*, 154, 21.
48. Carabineiro S.A.C., Bogdanchikova N., Avalos-Borja M., Pestryakov A., Tavares P.B., and Figueiredo J.L. (2011). *Nano Research*, 4, 180.
49. Carabineiro S.A.C., Bogdanchikova N., Pestryakov A., Tavares P.B., Fernandes L.S.G., and Figueiredo J.L. (2011). *Nanoscale Research Letters*, 6.
50. Carabineiro S.A.C., Bogdanchikova N., Tavares P.B., and Figueiredo J.L. (2012). *RSC Advances*, 2, 2957.
51. Carabineiro S.A.C., Silva A.M.T., Drazic G., Tavares P.B., and Figueiredo J.L. (2010). *Catalysis Today,* 154, 293.
52. Carabineiro S.A.C., Machado B.F., Drazic G., Bacsa R.R., Serp P., Figueiredo J.L., and Faria J.L. (2010). In *Scientific Bases for the Preparation of Heterogeneous Catalysts: Proceedings of the 10th International Symposium* (Gaigneaux E.M., Devillers M., Hermans S., Jacobs P.A., Martens J., and Ruiz A., eds), Vol. 175, p. 629.
53. Bond G.C. and Thompson D.T. (1999). *Catalysis Reviews: Science and Engineering*, 41, 319.
54. Carabineiro S.A.C. and Thompson D.T. (2007). In *Nanocatalysis* (Heiz E.U. and Landman U., eds), Springer-Verlag, New York, p. 377.
55. Carabineiro S.A.C. and Thompson D.T. (2010). In *Gold: Science and Applications* (Corti C. and Holliday R., eds), CRC Press, Taylor and Francis Group, New York, p. 89.
56. Jenkins R. and Snyder R.L. (1996). *Introduction to X-ray Powder Diffractometry*, John Wiley and Sons Inc, New York.
57. Silva C.G., Jurez R., Marino T., Molinari R., and Garca H. (2011). *Journal of the American Chemical Society*, 133, 595.
58. Bond G.C., Louis C., and Thompson D.T. (2006). *Catalysis by Gold*, Imperial College Press, London.
59. Ohno T., Sarukawa K., Tokieda K., and Matsumura M. (2001). *Journal of Catalysis*, 203, 82.

60. Parra S., Olivero J., Pacheco L., and Pulgarin C. (2003). *Applied Catalysis B: Environmental*, 43, 293.
61. Apolinário Â.C., Silva A.M.T., Machado B.F., Gomes H.T., Araújo P.P., Figueiredo J.L., and Faria J.L. (2008). *Applied Catalysis B: Environmental*, 84, 75.
62. Gokel G.W. (2004). *Dean's Handbook of Organic Chemstry*, 2nd edn, MacGraw-Hill, New York.

Index

AC, *see* activated carbon
acid hydrolysis 74–7, 79
activated carbon (AC) 3, 62–9, 71–4, 119, 150, 222, 224, 238
adsorbate 64, 72, 111–12, 119
adsorbent materials 69, 71
adsorption 10, 66, 69, 110–12, 119
adsorption capacities 9–10, 73–4, 152, 154
AFM, *see* atomic force microscopy
agricultural wastes 2, 68–70
amorphous cellulose 76
anatase 98, 153–4, 248–51, 253–6, 261
anode 176–7, 179–80
anode materials 178–9, 181, 183
antibacterial properties 136, 145, 153, 155–6
antimicrobial activity 222–5, 227, 237, 239, 241–2
antimicrobial applications 223–5
antimicrobial properties 144–5
antimony 179–80
aquatic environment 148, 221, 223
aqueous solutions 4, 10, 31, 38–9, 64, 67, 69, 72–3, 98, 111, 115, 136, 169, 197, 202
arsenic 2, 97–8, 149–50, 155
arsenic removal 98, 149
atomic force microscopy (AFM) 134–5, 207

bentonite 4, 6, 8, 65, 99, 110, 112–15

bentonite clay 110, 112
biodegradable nanocomposites 125, 137
biofibers 57
biological metabolism of wastewater 62
biopolymers 3, 14, 25, 57, 67
biosorption 69
brookite 248–9
bulk polymerization 129

carbon 48–9, 52, 54, 66–7, 152, 193, 263
carbon nanoparticles 64, 200
carbon nanotubes 64, 79, 193, 196, 201, 223
catalysts 172, 178, 196, 199, 223, 255, 259–60, 262, 264–5
CCD, *see* cold corona discharge
cell wall 50–1, 54, 75, 78
cellulose 3, 29, 36, 48–50, 54–7, 69, 73–9, 193, 199, 201
  microcrystalline 78–9
  microfibrillated 77–9
  nanocrystalline 47, 75, 77, 80
cellulose chains 54, 74–5
cellulose content 48, 55
cellulose fibers 56, 74, 76–7
cellulose material 48, 75, 78
cellulose microfibrils 50, 76, 78–9
cellulose nanocomposite materials 73
cellulose nanocomposites 73
cellulose nanocrystals 73, 75, 77–9
cellulose nanofibers 48, 73, 76
cellulose nanoparticles 77–8, 80

cellulose nanowhiskers 47, 77–8
cellulosic materials 75
cellulosic nanocomposites 47–8,
    50, 52, 54, 56, 58, 60, 62,
    64, 66, 68, 70, 72, 74, 76
cellulosic nanocomposites for
    treatment of wastewater
    47–8, 50, 52, 54, 56, 58,
    60, 62, 64, 66, 68, 70, 72,
    74, 76
ceramic materials 196
chemical contaminations 38, 144,
    146
chemical oxidative polymerization
    39, 225–6
chemical vapor deposition
    techniques 148, 154
chemisorption 109, 211–12
chitosan 1–8, 10–14, 25, 34, 64,
    67, 148, 223
chitosan-based polymer
    nanocomposites 1, 4–5,
    7, 9
chitosan-based polymer
    nanocomposites for heavy
    metal removal 2, 4, 6, 8,
    10, 12, 14
chitosan-capped gold
    nanocomposite 8
chitosan clay nanocomposite 4
chitosan-coated fly ash 6, 8
chitosan nanocomposite 4
    graphene oxide 12
chitosan nanoparticles 4
chitosan- magnetite
    nanocomposites 5
chitosan–clay nanocomposite 4
chromium 2, 58, 73, 97, 201
chromium removal 73, 201
clay-based nanocomposites 98
clay layers 102
clay minerals 98–9, 110

clay particles 98–9, 104–5,
    108–10, 114, 119, 128
clean water 118, 127, 143–4, 222
coacervation process, complex 38
coconut shell 67, 69, 71–2
cold corona discharge (CCD)
    174–5
coliform bacteria 223, 230, 237–8
    fecal 221–3
    isolating 222, 225, 228
coliforms 228, 230, 242
color removal 169, 184–5
composite materials 49, 51, 73,
    192, 209
    properties of 77, 192
composites 5–8, 47–9, 52–3, 75,
    98, 104, 112–13, 119, 126,
    128, 137, 192–3, 196, 214,
    232
    natural fiber 53
    reinforced 49
conjugated polymers 225
conventional nanocomposites 128
conventional wastewater
    treatment 58
copper 2, 58, 98
    removal of 12
core–shell nanocomposites 212,
    214
corona discharge method 174,
    176
cotton 48, 52, 54, 74–5, 78

DBPs, see disinfection by-products
dendrimers 134
dimensionally stable anode (DSA)
    179
dimensions, crystallite 254–6
diphenhydramine 248, 250
disinfection by-products 223
disinfection by-products (DBPs)
    144–5, 223

dissolved ozone 183
drinking water quality 208
DSA, *see* dimensionally stable anode
dye-containing wastewater 155
dye removal 63, 69

electrochemical ozone production (EOP) 167–8, 170, 172, 174, 176, 178–80, 182, 184
electrode materials 178–9
electrospinning 129–30
EOP, *see* electrochemical ozone production
Escherichia coli 135, 147, 221, 223
ethanol 197–8, 200, 233, 250, 252
ethylene vinyl acetate (EVA) 103, 110, 112–15
EVA, *see* ethylene vinyl acetate
EVA-bentonite nanocomposites 97–8, 100, 102, 104, 106, 108, 110, 112, 114, 116, 118

fiber sources 51
fibers 48–57, 75–7, 79–80, 119, 126, 199, 225, 234, 240
  cellulosic 48–9, 80
  mineral 48
  nanocomposite-based 243
flocculation 59, 61, 63, 144, 208
food industry 25, 29

gold 179, 200, 203, 209, 211, 249, 256, 260, 264–5
graft copolymerization 26, 30–1, 36, 39, 80
graphene oxide–chitosan nanocomposite 9
graphene polypyrrole nanocomposite 137

groundwater 149, 208
  contaminated 149
guar gum 25, 30–1
gum arabic 26, 33
gum gellan 29–30
gum ghatti 32–3
gum karaya 26–7
gum-polysaccharide-based nanocomposites 23–4, 26, 28, 30, 32, 34, 36, 38, 40
gum polysaccharides 25, 29, 31
gum xanthan 25, 28–9

Hammett constant 263–5
heavy metal, detection 191–2
heavy metal adsorption 4, 13–14, 109, 191
heavy metal ions 2, 4, 23, 38–9, 150, 191, 201–2, 210–12, 214
  detection 210–11
heavy metal removal 1–2, 4, 6, 8–14, 97–8, 100, 102, 104, 106, 108, 110, 112, 114, 116, 118
heavy metals 2, 4, 58, 67, 69, 97–8, 100, 108–10, 114–16, 118–19, 147, 149–51, 153–4, 156, 191
hemicelluloses 48, 50–1, 54–7, 69, 74–6
hexavalent chromium 4
high resolution transmission electron microscopy (HRTEM) 236, 242, 252, 256
HRTEM, *see* high resolution transmission electron microscopy
hydrogen peroxide 168, 170, 172
hydrophilic polymers 99

hydrophobicity 134, 136
hydroxyl radicals 171–2

industrial effluents 23–4, 26, 28, 30, 32, 34, 36, 38–40, 109, 125, 128, 135, 137
inorganic compounds 58, 169–71
intercalated nanocomposites 106–8, 128, 194
intercalation 101–4, 130
intraparticle diffusion 111–12
ion exchange 2, 12, 59, 63–4, 98–9, 109, 208
ions
 exchangeable 99–100
 heavy-metal 111
 toxic 64, 210–11

Langmuir isotherm 9–10
LCST, see lower critical solution temperature
liquid phase reductive deposition (LPRD) 251, 256–61, 264–5
localized surface plasmon resonance (LSPR) 209
locust bean gum 31
low-cost adsorbents 3, 14, 64–5
lower critical solution temperature (LCST) 34, 38
LPRD, see liquid phase reductive deposition
LSPR, see localized surface plasmon resonance

magnetic chitosan 4–5
magnetic chitosan nanocomposites 12
magnetic chitosan nanoparticles 5
magnetic Fe 150–1
magnetic graphene nanocomposites 202

magnetic nanoparticles 150, 152, 197
 synthesis of 151
magnetic particles 152
matrix materials 56–7
MCMs, see Mobil crystalline materials
melt-blending method 101, 103, 110, 112, 115–16, 118–19
melt intercalation 103
membrane filtration method 222, 225, 227–9
membrane separations 61, 63
membranes, nanocomposite-based 222, 227, 242
metal electroplating wastewaters 136
metal ions 3–4, 9–10, 12, 14, 39, 66–7, 135, 147, 151, 192, 201, 208–9, 214
metal nanoparticles 2, 4, 12, 209–10, 224
methylene blue 71, 128
microbial contaminations 144
microfibrils 50, 57, 74–6, 78
microorganisms 14, 25, 28–9, 59, 62, 171, 224
Mobil crystalline materials (MCMs) 199

nanocomposite applications 77
nanocomposite brittleness 118
nanocomposite materials 75, 126, 132, 134, 167
 functional 1
 hybrid 193
nanocomposite particles 111
nanocomposites
 biopolymer-based 24–5
 chitosan-based 1, 5, 12–14, 38
 chitosan–magnetite 4
 core–shell 191, 201–2, 206, 213

exfoliated 105, 107, 128–9, 194
glucomannan–chitosan-based 38
hydroxyapatite/chitosan 12
modified chitosan–montmorillonite 4
nylon 6–clay 102
pH-Responsive 35–6
polymer–clay 98, 116
polypropylene–clay 108
polysaccharide-based 24, 36, 38–9
stimuli-responsive 33, 35
synthesized 36
temperature-responsive 33–4
thermo-sensitive 34
nanofillers 193, 195–7
nanomaterials
  carbonaceous 134, 136
  natural 198–9
  synthetic 198–9
nanoparticles, magnetite 4, 10–12
nanoparticles for water purification 143–56
natural fiber surfaces 49
natural fibers 47–55, 57, 69
natural polysaccharides 24
natural waters 223, 250

OER, see oxygen evolution reaction
organic compounds 38, 47, 61–2, 172, 184, 201
organic modifier 101–3
organic pollutants 67, 144, 149–50, 152–3, 167–8, 170, 172, 174, 176, 178, 180, 182, 184
organic wastewater 131
organoclay 99–100, 104–5, 125
organoclay nanocomposites 132
oxidants 168–9, 172, 184, 226

oxygen evolution reaction (OER) 178–9
ozonation 59, 144, 169–73, 184
ozone 144, 168–76, 178–80, 183–5, 223
  molecular 169–70
ozone concentrations 174, 183–4
ozone decomposition 170–2
ozone generation, electrochemical 176, 178
ozone production 167, 173–5, 177–9
ozone water 172–3, 178

PCNs, see polymer–clay nanocomposites
pectin 48, 51, 54–5, 57, 74
PEM, see polymer electrolyte membrane
perchlorate removal 127–8
perchlorates 125, 127–8
pesticides 58, 68, 147–8, 152, 155
phenolic compounds 248, 250, 253, 261–3, 265
phenols, substituted 250, 263–5
plant fibers 51, 53–4, 71, 74
platelets 99, 104, 108, 128–9
polyacids 35
polyaniline nanocomposites 125, 127
polymer chains 24, 31, 33, 35, 101–2, 128–9, 194
polymer coacervation process 37
polymer electrolyte membrane (PEM) 176, 178
polymer nanocomposites 14, 125–6, 128–30, 132–6, 224, 227
polymer–clay nanocomposites (PCNs) 98–9, 103–5, 108–17

polymeric nanocomposites 98–9, 101, 103, 105, 107, 118
polymers, synthetic 25, 36, 40, 57
polysaccharides 25, 31, 39–40, 55, 69, 147, 199
polyurethane 114–15
PPY-Ag nanocomposites 226–8, 230–8, 241–2

QDs, *see* quantum dots
quantum dots (QDs) 199, 202

radical scavengers 171–2, 260
reverse osmosis 2, 62, 64, 134
rutile 248–51, 253–6, 261

scanning electron microscope (SEM) 10, 108, 133, 180, 182, 185, 191, 206, 222, 225, 242
scanning electron microscopy 133, 180, 206, 222, 225
scanning probe characterization technique 206–7
SEM, *see* scanning electron microscope
silica 69, 195, 201, 203–4
silica particles 197, 203
silicate layers 107, 194
silicates 105–6, 108, 110
silver nanocomposites 211
  oil-based polymer 132
silver nanoparticle nanocomposites 148
silver nanoparticles 145–8, 156, 200, 203, 224–6, 230–4, 236–42
  synthesis of 145–7
SiO$_2$ nanoparticles 197
smart nanocomposites 208–9, 211, 213
smectite clay 99

sodium borohydride 145, 148, 200
sulfate removal 137
sulfates 64, 125, 127, 148, 150
sulfide 127, 169
surface plasmon resonance 203–4
suspension polymerization 37

TEM, *see* transmission electron microscopy
transmission electron microscopy (TEM) 105, 108, 132, 134–5, 191, 206–7, 222, 225, 236, 252

UV-visible spectral studies 11–12

wastewater
  contaminated 170
  decolorizing dyestuff 184
  domestic 58
  electroplating 73
  industrial 58, 64, 67, 250
  reclamation of 127
  synthetic 72
wastewater contaminant 127
wastewater handling 48, 68
wastewater reclamation 127
wastewater treatment
  large-scale 151
  organic 131
wastewater treatment adsorbents 65
water
  contaminated 38, 98, 144, 149, 153, 170, 222–3
  contamination of 221, 223
  deionized 4, 197
water disinfection 148, 224
water pollution 23, 151, 191
water purification 68, 134, 143–56, 191, 223, 249

waterborne diseases 143–4, 222
wood fibers 48, 75–6

X-ray diffraction (XRD) 105–6, 108, 129–31, 180, 231, 242, 251, 256

xanthan gum 36
XRD, *see* X-ray diffraction

zeolites 64, 134–5, 194, 199
zero-valent iron 144, 148–50, 156
zinc 58, 72, 210